Wireless Networks

Series Editor

Xuemin Sherman Shen, University of Waterloo, Waterloo, Canada

The purpose of Springer's Wireless Networks book series is to establish the state of the art and set the course for future research and development in wireless communication networks. The scope of this series includes not only all aspects of wireless networks (including cellular networks, WiFi, sensor networks, and vehicular networks), but related areas such as cloud computing and big data. The series serves as a central source of references for wireless networks research and development. It aims to publish thorough and cohesive overviews on specific topics in wireless networks, as well as works that are larger in scope than survey articles and that contain more detailed background information. The series also provides coverage of advanced and timely topics worthy of monographs, contributed volumes, textbooks and handbooks.

Zhirun Zheng • Zhetao Li • Xuemin Shen

Perturbation Based Privacy in Crowdsensing

Springer

Zhirun Zheng
Suwon, Korea (Republic of)

Zhetao Li
Guangdong, China

Xuemin Shen
Waterloo, ON, Canada

ISSN 2366-1186 ISSN 2366-1445 (electronic)
Wireless Networks
ISBN 978-3-031-95051-3 ISBN 978-3-031-95052-0 (eBook)
https://doi.org/10.1007/978-3-031-95052-0

This Springer imprint is published by the registered company Springer Nature Switzerland AG
The registered company address is: Gewerbestrasse 11, 6330 Cham, Switzerland

Preface

With the rapid development of communication networks and mobile devices, crowdsensing has emerged as a popular and widely adopted information collection paradigm, in which mobile devices are recruited by a crowdsensing server to act as intelligent sensors that collect and sense various environmental data, such as temperature and humidity. Although crowdsensing has already been applied in various areas such as intelligent transportation and urban computing, to provide high-quality services, it also raises serious privacy concerns because the sensing data contains a large amount of sensitive information. As a result, privacy preservation in crowdsensing is crucial, especially under information security laws and regulations, such as the European General Data Protection Regulation. To address these privacy concerns in crowdsensing systems, perturbation-based privacy-preserving techniques have been widely deployed. However, noise perturbation added by the techniques could inevitably disrupt the inherent features of data, leading to subpar service quality and privacy preservation. In addition, attackers can utilize the spatiotemporal feature of trajectories to identify that such movement is unfeasible, and subsequently launch more powerful privacy attacks. Furthermore, attackers can also subtly exploit the noise perturbation to conceal their malicious behavior, thereby stealthily tampering with the reported information and subverting crowdsensing results.

In this monograph, we focus on perturbation-based privacy in crowdsensing. In Chap. 1, we present an explicit overview of crowdsensing systems and privacy challenges, and briefly discuss how the noise added by perturbation-based privacy-preserving techniques could inevitably degrade data quality and facilitate the success of data poisoning attacks on crowdsensing. In Chap. 2, we give a comprehensive review of classical privacy notions for perturbation-based privacy-preserving techniques, and theoretically analyze the relations between these privacy notions. Powered by these privacy notions and theorems, we conduct a series of studies on privacy preservation in crowdsensing systems from three dimensions of data privacy, data utility, and data poisoning. Specifically, in Chaps. 3 and 4, we investigate the trade-off between data privacy and data utility (a.k.a. data privacy vs. data utility), aiming to protect users' data privacy while providing high-quality services. The privacy-preserving techniques from Chap. 3 can protect users' data and semantic privacy while preserving semantic-aware data utility, whereas the techniques in Chap. 4 can provide users with high-quality dynamic pricing and ride-matching services while protecting their data privacy. Additionally, in Chaps. 5 and 6, we study the trade-off between data privacy and data poisoning (a.k.a. data privacy vs. data poisoning), striving to protect users' data privacy while defending against data poisoning attacks concealed within noise perturbation. The defense techniques from Chap. 5 can defend against data poisoning attacks concealed within the noise perturbation introduced by local differential privacy protocols, while the defense techniques in Chap. 6 can counter the data poisoning attacks hidden behind the noise perturbation added by central differential privacy protocols. Through these defense techniques, we can mitigate the threats posed by perturbation-based privacy-preserving techniques in crowdsensing. Finally, we conclude the monograph and discuss future research directions for perturbation-based privacy preservation in crowdsensing systems.

We would like to thank Prof. Chao Chen, Prof. Bin Guo, Prof. Suiming Guo, Prof. Cheng Huang, Prof. Hongbo Jiang, Prof. Dongxiao Liu, Prof. Mushu Li, Prof. Saiqin Long, Prof. Kui Ren, Dr. Dengbiao Tu, Prof. Fu Xiao, Prof. Ke Xu, and Dr. Leo Yu Zhang for their valuable contributions to this monograph. We would also like to thank all members of the Broadband Communications Research (BBCR) group at the University of Waterloo for their valuable suggestions and support.

Suwon, Korea (Republic of) Zhirun Zheng
Guangdong, China Zhetao Li
Waterloo, ON, Canada Xuemin (Sherman) Shen

Declarations

Competing Interests The authors have no competing interests to declare that are relevant to the content of this manuscript.

Contents

1 **Introduction** 1
1.1 An Overview of Crowdsensing 1
1.2 Privacy Challenges in Crowdsensing 6
1.3 Aims and Organization of This Monograph 13
References 15

2 **Perturbation-Based Privacy Preservation** 21
2.1 Classical Privacy Notions 21
2.2 Relations Between Privacy Notions 24
2.3 Summary 35
References 35

3 **Semantic-Aware Trajectory Privacy Preservation in Crowdsensing** 37
3.1 Introduction 37
3.2 Problem Statement and Basic Concepts 39
3.3 Privacy and Utility Metrics 42
3.4 Semantic-Aware Privacy Mapping Mechanism 47
3.5 Privacy Analysis 57
3.6 Performance Evaluation 60
3.7 Summary and Further Reading 68
References 68

4 **Pricing-Aware Location Privacy Preservation in Crowdsensing** 71
4.1 Introduction 71
4.2 Problem Statement and Basic Concepts 74
4.3 Utility Loss Metrics 76
4.4 Pricing-Aware Privacy Mapping Mechanism 82
4.5 Privacy Analysis 92
4.6 Performance Evaluation 93
4.7 Summary and Further Reading 100
References 101

5 Data Poisoning Attacks and Defenses to LDP-Based Crowdsensing ... 103
5.1 Introduction ... 103
5.2 Problem Statement and Basic Concepts ... 105
5.3 Data Poisoning Attacks Hidden Behind the LDP Noise ... 111
5.4 Countermeasures: Designing Optimal Defenses ... 121
5.5 Computational Complexity and Limitations ... 125
5.6 Performance Evaluation ... 126
5.7 Summary and Further Reading ... 134
References ... 134

6 Data Poisoning Attacks and Defenses to CDP-Based Crowdsensing ... 137
6.1 Introduction ... 137
6.2 Problem Statement and Basic Concepts ... 140
6.3 Formulating Game Model Between Attacks and Defenses ... 146
6.4 Finding Optimal Data Poisoning Attacks and Defenses ... 148
6.5 Computational Complexity of Attacks and Defenses ... 162
6.6 Performance Evaluation ... 162
6.7 Summary and Further Reading ... 173
References ... 174

7 Conclusions and Future Research Directions ... 177
7.1 Conclusions ... 177
7.2 Future Research Directions ... 179

Acronyms

ABC	Artificial Bee Colony
ABIE	Average Bayesian Inference Error
ADMM	Alternating Direction Multiplier Method
AG	Attack Gain
APME	Average Price Multiplier Error
ASDRE	Average Supply-Demand Relationship Error
AUD	Average Utility Damage
CDP	Central Differential Privacy
CRH	Conflict Resolution on Heterogeneous
DE	Differential Evolution
DP	Differential Privacy
EMD	Earth Mover's Distance
EU	European Union
GDPR	General Data Protection Regulation
ITP	Information-Theoretic Privacy
JS	Jensen-Shannon
KL	Kullback-Leibler
LBS	Location-Based Services
LDP	Local Differential Privacy
ODA	Optimal Defenses Algorithm
PADP	Pricing-Aware Differentially Private
PoI	Point of Interest
PSO	Particle Swarm Optimization
RHS	Right-Hand Side
RMUM	Ride-Matching Utility Metric
ROC	Receiver Operating Characteristic
RoD	Ride-on-Demand
SEITP	SEmantic-aware Information-Theoretic Privacy
USD	United States Dollar

Chapter 1
Introduction

1.1 An Overview of Crowdsensing

With the development of global digital economies, mobile devices (e.g., smartphones and smartwatches) have become indispensable to our daily activities. As reported in [1] and [2], 5.6 billion users (approximately 69% of the global population) subscribed to mobile services by the end of 2023, and the mobile devices market is expected to reach USD 493.26 billion by 2029. Furthermore, the global crowd analytics market is predicted to reach USD 5.3 billion by 2030, from USD 1.69 billion in 2024 at a compound annual growth rate of 20.68% [3]. Under these circumstances, crowdsensing systems have great potential for widespread deployment, leveraging crowd intelligence to improve citizens' daily lives.

1.1.1 Evolutionary Path of Crowdsensing

Crowdsensing is a new large-scale sensing paradigm derived from the concept of crowdsourcing [4], which was first introduced by Ganti et al. [5] to sense and collect environmental data, such as temperature and humidity. In 2005, James Surowiecki [6] revealed in his book titled The Wisdom of Crowds a general phenomenon: the aggregation of data or information from a group of people often results in better decisions than those made by a single person from the group. Powered by this phenomenon, Jeff Howeand and Mark Robinson [4] coined the term crowdsourcing in 2006, defining it as the practice of obtaining needed services or content by soliciting contributions from a large group of people and especially from the online community rather than from traditional employees or suppliers.

Z. Zheng et al., *Perturbation Based Privacy in Crowdsensing*, Wireless Networks,
https://doi.org/10.1007/978-3-031-95052-0_1

A typical application of crowdsourcing is Wikipedia,[1] where tens of thousands of contributors collaboratively create the largest encyclopedia of the world [7].

Compared to crowdsourcing, the term crowdsensing was created by Ganti et al. [5] in 2011 by combining opportunistic sensing and crowd intelligence, emphasizing more autonomous sensing and minimal user involvement (e.g., continuous location sharing in Google Maps[2]). In 2014, Guo et al. [8] provided a formal definition of crowdsensing as follows: a new sensing paradigm that empowers ordinary citizens to contribute data sensed or generated from their mobile devices, and aggregates and fuses the data in the cloud for crowd intelligence extraction and human-centric service delivery. After that, researchers in both academia and industry have conducted a series of studies [7, 9] and developed a range of applications, including CrowdOS[3] and CrowdHMT.[4] In recent years, the rapid development of 5G communication networks and mobile devices has ushered crowdsensing into a new era [10].

1.1.2 Architecture of Crowdsensing

As a popular and widely adopted sensing paradigm, crowdsensing systems utilize sensors (e.g., accelerometers and gyroscopes) embedded in mobile devices to collect various environmental data [11, 12]. Figure 1.1 shows the architecture of crowdsensing systems, consisting of three layers: a user layer, a platform layer, and a sensing layer.

- **User Layer:** This layer involves sensing task requesters, including individual requesters (e.g., drivers and doctors) and non-individual requesters (e.g., government agencies and industrial factories), who upload tasks to the platform layer to obtain the corresponding sensing data. For example, drivers collect traffic information (e.g., congestion and pothole data) to enhance their driving experience [13], and government agencies promptly identify and address infrastructure issues by monitoring conditions of bridges and tunnels [14].
- **Platform Layer:** This layer gathers sensing data by performing the following four components: worker recruitment, task allocation, data collection, and data aggregation. These four components are strictly executed within a crowdsensing server, and their detailed procedure is outlined below.
  - *Worker Recruitment:* This component publicly recruits individuals with sensor-rich mobile devices to engage in sensing tasks. In particular, recruiting a sufficient number of qualified workers is crucial for the success of

[1] https://www.wikipedia.org/.

[2] https://www.google.com/maps/.

[3] https://www.crowdos.cn/.

[4] http://www.crowdhmt.com/.

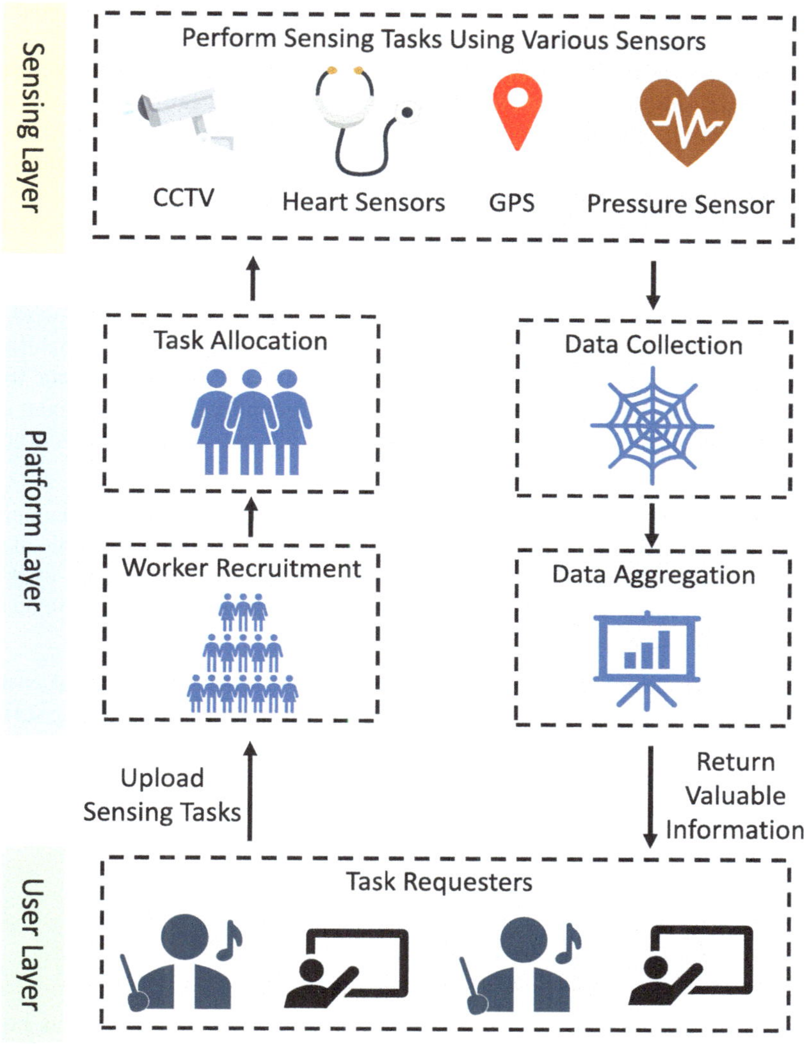

Fig. 1.1 Crowdsensing architecture

crowdsensing systems [15]. For example, workers must be recruited from various locations to photograph water bodies in water quality monitoring tasks, ensuring high-accuracy monitoring results [16].

– *Task Allocation:* This component analyzes sensing task requirements (e.g. location, time, and sensor quality) and then assigns the task to specific workers accordingly. For instance, the mobile devices of participating workers must be equipped with high-quality microphones, creating fine-grained noise maps in noise pollution measurement tasks [17].

- *Data Collection:* This component collects sensing data shared by workers and then stores the data on the crowdsensing server for aggregating and analyzing. For example, the server gathers traffic conditions information (e.g., traffic volume and speed) for individual travel planning, thereby improving their travel experience [18].
- *Data Aggregation:* This component extracts valuable information from the collected sensory data. Although massive amounts of data are collected in the data collection component, obstacles such as quality variation and multimodal hinder its direct use. Specifically, the data from different workers have distinct quality and credibility due to varying sensing capabilities (e.g., high-resolution and low-resolution cameras). Besides, the data from various types of sensors (e.g., cameras and microphones) are multimodal and heterogeneous. As a result, aggregating the collected sensing data is crucial for crowdsensing systems.

- **Sensing Layer:** This layer, consisting of various sensors such as cameras, microphones, and GPS, enables workers to perform sensing tasks and upload the corresponding sensing data. For example, in public transportation tasks, workers use radar and cameras to sense the traffic conditions around them and upload the conditions along with their location from GPS [13].

Based on these three layers, we present the detailed workflow of crowdsensing systems as follows. The requesters submit sensing tasks to the crowdsensing server, aiming to obtain valuable information for improving service quality. Then, the server assigns the tasks to a group of workers and asks them to upload the corresponding sensing data. Finally, the server collects sensing data shared by workers, aggregates it, and returns the aggregated results to requesters.

1.1.3 Applications of Crowdsensing

With the ubiquitous usage of sensor-rich mobile devices, rising to various novel crowdsensing applications. Next, we briefly present and discuss some representative applications, as shown in Fig. 1.2.

- **Environment Monitoring:** Environment pollution has become a global issue, deeply affecting people's daily lives. To monitor environmental pollution levels, various crowd-powered environment monitoring applications have been developed [17, 19, 20]. For example, NoiseTube [20] could measure personal exposure to environmental noise using microphones and create city noise maps accordingly. U-Air [19] could infer fine-grained air pollution levels (e.g., PM2.5) using heterogeneous sensing data, including the air quality data sensed by

Fig. 1.2 Crowdsensing applications

existing monitoring stations and the traffic information and points of interest (PoIs) reported by GPS.

- **Urban Planning:** Well-planned urban can improve the quality of citizens' lives regarding convenience, comfort, safety, and security. However, collecting data for urban planning is an increasingly important challenge. In recent years, the emergence of crowdsensing has facilitated large-scale urban data collection, driving the growth of crowd-enabled urban planning applications. For instance, Geo-Spotting [21] could determine the optimal placement for new retail stores by analyzing geographic and human mobility patterns. Tang et al. [22] and Wang et al. [23] respectively study the community detection problem and the community discovery and profiling issues by analyzing the corresponding sensing data.
- **Social Recommendation:** With the abundant data gathered from crowdsensing systems, various social recommendation services can be enhanced, including

place recommendation, itinerary planning, and activity suggestion. For example, Yang et al. [24] provide users with personalized place recommendation services by analyzing their check-in data and comments shared on FourSquare.[5] Yu et al. [25] develop a travel package recommendation system that provides personalized travel packages by considering user preferences, PoI characteristics, and spatiotemporal constraints.

- **Public Safety:** Public safety refers to detecting or preventing social or natural events (e.g., crimes and disasters) that may endanger the safety of citizens. Recently, crowdsensing has been utilized to assist in crime prevention and disaster relief. For instance, Ballesteros et al. [26] reveal crimes by analyzing the data collected by geo-social networks and then propose iSafe accordingly, a crowd-powered public safety system designed to evaluate the safety of users. Sakaki et al. [27] could give a near-real-time report of earthquakes by analyzing large-scale mobile phone user data on pre-earthquake and post-earthquake movement behaviors.
- **Transportation:** Crowdsensing injects new vitality into intelligent transportation systems, including real-time traffic forecasting and road condition monitoring, intending to enhance citizens' travel experiences and contribute to energy conservation and emissions reduction. For example, PotHole [28] could identify holes in streets using vibration and location data collected from smartphones. SmartTrace [18] helps identify mobility patterns or a given trajectory's popularity for transit planning.
- **Healthcare:** In recent years, healthcare has become an increasingly important challenge with the growing aging population. Crowd-enabled healthcare applications could continuously monitor individuals' health status through sensor-rich wearable devices (e.g., smartwatches), and then improve healthcare delivery according to the collected health data. For instance, in public health monitoring, Ferguson et al. [29] estimates the level of influenza-like illnesses by analyzing health-related search queries. In personal well-being management, DietSense [30] allows people to photograph and share their dietary choices, and get suggestions from online experts for weight loss.

1.2 Privacy Challenges in Crowdsensing

In recent years, privacy protection has gained attention from individuals and governments. From the individual view, a survey report by the global information security association ISACA revealed that approximately 90% of users expressed concerns about the leakage of their sensitive information, while only 6% indicated that they were not particularly concerned about their privacy [31]. From the government view,

[5] https://foursquare.com/.

the European Union (EU) enacted the General Data Protection Regulation[6] (GDPR) in 2018, and China subsequently implemented the personal information protection law[7] and the data security law.[8] These laws are designed to regulate the collection and use of personal information by organizations and companies, imposing severe penalties for violations, e.g., Facebook's fine of 91 million euros for violating the GDPR [32]. Nevertheless, privacy breaches continue to occur frequently. For example, the automaker NIO's user information and vehicle sales data were stolen in 2022, accompanied by a ransom demand of $2.25 million [33]. Moreover, leaks of users' sensing data could expose personal sensitive information, and even be exploited by attackers to launch data poisoning attacks that subvert the crowdsensing outputs [34–36]. These concerns hinder users from participating in crowdsensing systems, resulting in substantial financial losses and potential failure.

1.2.1 Motivations: Why Privacy Matters

Although crowdsensing greatly improves daily convenience for citizens, it also poses serious risks to personal privacy [34–39]. In particular, the data collection component could pose significant personal privacy risks related to identity, health status, and social relationships [38, 39]. Additionally, the user recruitment component is highly susceptible to exploitation by attackers, who launch data poisoning attacks to overturn the crowdsensing results (i.e., outputs of the data aggregation component) [34–36]. Next, we explain why privacy matters in crowdsensing systems from two aspects: the personal information leaks caused by sharing real sensing data and the poisoning threats arising from perturbation-based privacy-preserving mechanisms.

The data collection component directly gathers users' real sensing data, causing significant privacy leakage risks. This is because the sensing data contains a large amount of sensitive information, such as the location data from GPS could reveal identity, social relationship, visit purpose, and religious beliefs [38–41]. Specifically, Montjoye et al. [40] pointed out that attackers could successfully identify 95% of individuals in an anonymized location trajectory dataset with knowledge of just four spatiotemporal location points, leading to personal identity privacy leaks. In social relationship privacy, attackers could expose users' social relationships from location trajectory data, since friends often meeting during holidays, such as dining out or traveling [38]. In visit purpose privacy, when attackers know that the majority of PoI near a specific location are hospitals, they could infer that the user's purpose for visiting this location is to go to hospital [39]. In addition, given the users' frequent mobility patterns, attackers could infer victims' religious

[6] https://gdpr.eu/.

[7] https://www.gov.cn/xinwen/2021-08/20/content_5632486.htm.

[8] https://www.gov.cn/xinwen/2021-06/11/content_5616919.htm.

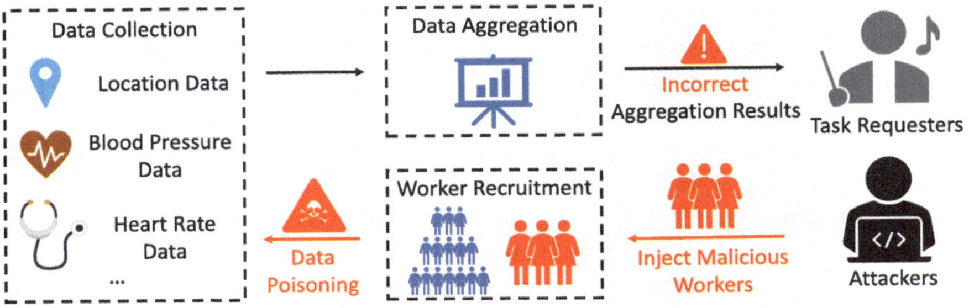

Fig. 1.3 Data poisoning attacks against crowdsensing systems

beliefs (i.e., visiting religious sites frequently) and political inclinations (i.e., visiting political party event centers frequently) from these mobility patterns [38, 39, 41].

As shown in Fig. 1.3, attackers can exploit the worker recruitment component to inject malicious workers into crowdsensing systems, launching data poisoning attacks to subvert the aggregation results [34–36]. The assumption that attackers could inject malicious workers into crowdsensing systems (i.e., the attackers' capabilities) is realistic. As reported by Thomas et al. [42], attackers can easily create and control large numbers of fake accounts on services like Twitter and Google. In addition, the motivation behind attackers' attempts to disrupt aggregation results (i.e., the attackers' goal) is meaningful. For example, attackers manipulate traffic conditions displayed on Waze by directing malicious workers to spread false traffic information, inducing victims to travel according to the attackers' intentions [43]. Furthermore, many state-of-the-art works have demonstrated that differential privacy (DP), as a provable perturbation-based privacy notion, is susceptible to data poisoning attacks [43, 44], intensifying concerns about deploying DP in crowdsensing systems. As a result, it is crucial to explore whether DP could pose poisoning threats to perturbation-based privacy-preserving crowdsensing systems.

1.2.2 Data Privacy vs. Data Utility

The sensing data contains abundant location information and may even be specific locations. For example, in road condition sensing tasks, workers share data that includes condition details (e.g., potholes and congestion) and location information. In air quality sensing tasks, the shared data consists of pollution levels (e.g., PM2.5) and location information. In traffic flow sensing tasks, the shared data consists of specific locations. However, the noise introduced by perturbation-based privacy-preserving mechanisms could degrade the quality of essential services in crowdsensing systems, such as dynamic pricing and ride-matching services. As a result, we review state-of-the-art privacy-preserving mechanisms in crowdsensing systems, focusing on balancing privacy protection and service quality (i.e., data privacy vs. data utility).

Location Trajectory Privacy Existing location privacy protection techniques in crowdsensing systems are generally based on DP and information-theoretic privacy (ITP). In the DP-based location privacy protection techniques, Andrés et al. [45] proposed the privacy notion of geo-indistinguishability based on DP and then a perturbation-based privacy-preserving mechanism is designed to ensure that each location in a trajectory is indistinguishable from its perturbed counterparts. Sun and Tay [46] studied the relationships between various privacy notions (e.g., ε-average information privacy, ε-mutual information privacy, ε-information privacy, and ε-local differential privacy), and then proposed a privacy-preserving mechanism to protect users' data privacy based on ε-information privacy and ε-local differential privacy. Jiang et al. [47] proposed a robust location privacy protection algorithm, called RobLop, for preserving data privacy against location-dependent attacks (i.e., maximum movement boundary attacks and arrival boundary attacks) in continuous location-based services (LBS) queries. However, recent work in [48] has shown that DP is inapplicable for correlated data (e.g., location trajectory data) publishing due to the different correlations between noise and original data. In the ITP-based location privacy protection techniques, Li et al. [49] proposed a general mutual information-based ITP metric to protect individual privacy in the distributed scenario. Wang et al. [50] obtained an arbitrarily strong utility-privacy trade-off in the distributed scenario by studying the class of privacy techniques offered by linear compression and noise perturbation, and then proposed a privacy protection mechanism to find the maximum estimation privacy achievable without compromising the utility. Zhang et al. [51] proposed a trajectory-level data privacy metric based on mutual information. To obtain the bounds of the proposed privacy metric, the authors assumed that the real or perturbed trajectory follows a first-order Markov chain, i.e., the perturbed location may leak the data privacy of the current real location and the one-step past real location.

Dynamic Pricing Utility On-demand ride-hailing services, as representative applications of crowdsensing systems, can adaptively adjust supply and demand through dynamic pricing. In recent years, a lot of works have been done around dynamic pricing [52–58]. For example, Guo et al. [53] studied the process of searching for the best price before getting on a car, and the searching process reflects the riders' reaction to dynamic prices. Guo et al. [52] studied the effect of traditional counterparts (e.g., metro, bus, and taxi) on the dynamic prices, and proposed a dynamic price prediction approach based on multi-source urban data. Guo et al. [54] studied how drivers seek riders to earn more under the dynamic pricing mechanism, and proposed the RoD-Revenue to predict drivers' revenue based on multi-source urban data. Guo et al. [55] studied how riders maximize revenue under the dynamic pricing mechanism, and proposed the route recommendation approach based on multi-source urban data. Additionally, some works have investigated better dynamic pricing strategies from the view of economics [56–58]. For instance, Huang et al. [57] proposed the dynamic pricing scheme based on deep reinforcement learning to balance the demands and supplies better. Shao et al. [58] devised a novel dynamic (or fair) pricing policy under asymmetrical demand patterns. As a result, dynamic pricing plays a crucial role in the on-demand ride-hailing services, and is designed based on supply and demand. However, perturbation leads to severe dynamic pricing utility loss because it disrupts the supply-demand relationship within the service area.

Ride-Matching Utility Although ride-matching is essential in ride-hailing services (e.g., Uber[9] and DiDi[10]), it is seriously damaged by perturbation. Thus, a lot of works [59–64] have been proposed to protect users' location privacy while providing them with high-quality ride-matching service. For example, Wang et al. [59] used the average travel distance of selected drivers (calculated based on riders' real location and drivers' perturbed location) instead of real geo-distance in the ride-matching service. Then, finding a driver closest to the rider (i.e., ride-matching service) is formulated as a mixed-integer non-linear programming problem that minimizes selected drivers' travel distance to the rider's location given the constraint of differential privacy. Wang et al. [63] proposed a novel location obfuscation mechanism that is formulated as a single-object minimization problem and aims to minimize the utility loss caused by perturbation under the constraints of ε-differential privacy and δ-distortion privacy. Following these works, Wang et al. [64] proposed an optimal task allocation framework with the differential-and-distortion. However, estimating the real geo-distance based on known information, such as the observed locations and the privacy-preserving mechanism, is inaccurate, leading to lower-quality ride-matching service. To improve the quality of ride-matching service, some works [60–62] have designed various privacy-preserving mechanisms that do not change the ranking of real locations' relative distance. For example, To et al. [62] proposed a framework for assigning tasks (or riders) to workers (or drivers) in an online manner without compromising the location privacy of both workers and tasks. Particularly, they designed a novel privacy mechanism based on hierarchically well-separated trees to preserve the ranking of real locations' relative distance. Following this work, Wang et al. [60] propose a planar Laplace distribution-based privacy mechanism that does not change the ranking of these locations' relative distances, and then design a threshold-based online task assignment mechanism that could deal with the one-worker-many-tasks assignment. However, it is difficult to achieve high-quality ride-matching service by just keeping the ranking unchanged.

As a result, existing works on perturbation-based location privacy protection mechanisms mainly suffer from the following two main challenges:

- **How to protect both trajectory data privacy and semantic privacy?** Existing perturbation-based trajectory privacy protection mechanisms ignore the spatiotemporal correlations among locations within a trajectory and the causal correlation (a type of directed correlation) between the real and perturbed trajectories, leading to serious data privacy leakage. Furthermore, those mechanisms

[9] https://www.uber.com/.

[10] https://web.didiglobal.com/.

ignore the PoI information of specific locations, causing users' visit purpose (a.k.a. semantic privacy) leaks. Since data privacy and semantic privacy generally do not imply each other (i.e., protecting data privacy does not prevent, and may even exacerbate, semantic privacy leaks), protecting both types of privacy is challenging.

- **How to provide users with high-quality dynamic pricing and ride-matching services simultaneously?** Noise from perturbation-based privacy protection mechanisms can protect personal privacy but degrades dynamic pricing and ride-matching services. Existing works cannot sufficiently guarantee these services, yet they are essential for on-demand ride-hailing and are mutually influential, i.e., preserving one may harm the other. Therefore, designing a perturbation-based privacy protection mechanism that maintains high-quality dynamic pricing and ride-matching utility is nontrivial.

1.2.3 Data Privacy vs. Data Poisoning

Data poisoning attacks, also commonly referred to as fake data injection attacks, involve the deliberate manipulation of data (i.e., poisoning) to corrupt computational results derived from that data. These attacks can significantly impact the reliability of data-driven systems, such as crowdsensing. In recent years, with the frequent occurrence of privacy breaches, DP has been widely adopted in crowdsensing systems to protect sensing data privacy. However, state-of-the-art studies have revealed that DP provides provable privacy guarantees for users but remains inherently vulnerable to data poisoning attacks. Therefore, we investigate whether DP can facilitate the success of data poisoning attacks against crowdsensing systems (i.e., data privacy vs. data poisoning). Next, we review the data poisoning attacks and defenses to crowdsensing and DP.

Data Poisoning Attacks and Defenses to Crowdsensing Systems The truth discovery methods, a category of aggregation techniques extensively utilized in crowdsensing systems for extracting valuable information, could mitigate conflicting data, but well-crafted malicious inputs can still bypass them [34, 65–70]. For example, focusing on categorical (or discrete) data, Miao et al. [67] proposed intelligent data poisoning attacks to crowdsensing with conflict resolution on heterogeneous (CRH, a type of truth discovery methods), and Miao et al. [66] studied data poisoning attacks against crowdsensing empowered with Dawid-Skene model (a type of truth discovery methods). Additionally, focusing on continuous data, Fang et al. [65] proposed optimization-based data poisoning attacks against crowdsensing empowered with CRH and Gaussian truth model (a type of truth discovery methods), and Zhang et al. [69] studied multi-round data poisoning attacks to crowdsensing with TruthFinder (a type of truth discovery methods). In these attacks, malicious workers can masquerade as normal workers, tricking truth discovery methods.

In recent years, robust truth discovery methods [71, 72] are proposed to mitigate data poisoning attacks. For instance, Huang et al. [71] investigated the data poisoning attacks on crowdsensing with truth discovery methods, and proposed a robust approach to counter such attacks. This robust approach integrates the reliability evaluation and worker filtering processes into CRH, where the reliability evaluation process estimates the reliability of workers, and then the worker filtering process removes unreliable workers. Zhang et al. [72] studied robust truth discovery methods against multi-round data poisoning attacks in crowdsensing, including the detection of malicious workers. In addition, Fang et al. [65] aims to arm the crowdsensing with malicious workers detection capability, and proposed two defenses to reduce the impact of malicious workers, i.e., the median-of-weighted-average defense and the maximize influence of estimation defense.

Data Poisoning Attacks and Defenses to Differential Privacy Recent years have witnessed a remarkable surge in the field of data poisoning attacks against DP, highlighting the escalating attention and momentum in this critical research domain [43, 44, 73–77]. For example, Giraldo et al. [43] found optimal data poisoning attacks to mislead a classifier that detects anomalies, disguising false data into normal data by exploiting central differential privacy (CDP) noise. Cao et al. [44] focused on local differential privacy (LDP) protocols for frequency estimation and heavy hitter identification, which are two fundamental data analytics tasks. They put forth targeted data poisoning attacks with the objective of manipulating these LDP protocols to inaccurately estimate high frequencies for attacker-selected items or falsely identify them as heavy hitters. A concurrent and independent work [73] studied untargeted data poisoning attacks to noninteractive LDP protocols, which aim to degrade the overall performance of these protocols. Following these works, Wu et al. [74] directed their attention toward LDP protocols for key-value data, and proposed novel targeted data poisoning attacks. In such LDP protocols, an aggregator aims to simultaneously estimate the frequency and mean value of each key among the collected key-value data, where the key-value data is a potentially heterogeneous data type (i.e., keys are categorical and values are numerical). Particularly, they formulated the attacks as a two-objective optimization problem, aiming to simultaneously maximize the frequencies and mean values of some attacker-chosen target keys. Li et al. [75] focused on LDP protocols for mean and variance estimation, and proposed a fine-grained poisoning attack to fine-tune and simultaneously manipulate mean and variance estimations. Imola et al. [77] observed LDP makes graph degree estimation protocols more vulnerable to poisoning attacks. These studies strongly advocate caution when considering deploying DP, as it makes systems inherently vulnerable to data poisoning attacks. In addition, Imola et al. [77] focused on differentially private graph analysis, and designed robust degree estimation protocols under LDP that can significantly reduce the threats caused by LDP noise.

As a result, existing works on data poisoning attacks and defense techniques in privacy-preserving crowdsensing systems face two main challenges:

- **How to defend against poisoning threats arising from local differential privacy in crowdsensing systems?** LDP consists of perturbation and calibrating. Specifically, the calibrating operation can mitigate the bias caused by LDP noise, improving service quality significantly. However, the calibrating operation can amplify the impact of poisoning, posing new poisoning threats to crowdsensing systems. Therefore, defending against poisoning threats posed by LDP is crucial for its deployment in crowdsensing systems.
- **How to defend against poisoning threats arising from central differential privacy in crowdsensing systems?** Although CDP can protect users' sensing data privacy by adding noise, it is inherently vulnerable to data poisoning attacks. Besides, attackers can disguise their poisoning behaviors as CDP noise, evading detection by truth discovery methods. Thus, defending against poisoning attacks that hide behind CDP noise is crucial for CDP-based privacy-preserving crowdsensing systems.

1.3 Aims and Organization of This Monograph

The noise added by perturbation-based privacy protection mechanisms inevitably degrades data utility and even facilitates the success of data poisoning attacks against crowdsensing systems. Intuitively, better privacy preservation requires more noise perturbation, but excessive noise reduces data utility and enables more stealthy data poisoning attacks. Therefore, we investigate perturbation-based privacy in crowdsensing systems from three dimensions of data utility, data privacy, and data poisoning, aiming to protect users' data privacy while providing them high-quality data-driven services and mitigating poisoning threats arising from the noise perturbation.

1.3.1 Aims of This Monograph

As shown in Fig. 1.4, we focus on the trade-off between data utility and data privacy (i.e., data utility vs. data privacy), and aim to protect each user's privacy while providing them with high-quality services. In addition, we focus on the trade-off between data privacy and data poisoning (i.e., data privacy vs. data poisoning), striving to defend against data poisoning attacks hidden behind the noise perturbation. To this end, we have conducted the following research.

- **To protect users' sensing data privacy while providing high-quality services (i.e., data privacy vs. data utility),** we propose the semantic-aware trajectory privacy preservation and the pricing-aware location privacy preservation. These two privacy-preserving mechanisms solve the issues in existing works that fail to protect users' trajectory data and semantic privacy, as well as those that do not provide high-quality dynamic pricing and ride-matching services.

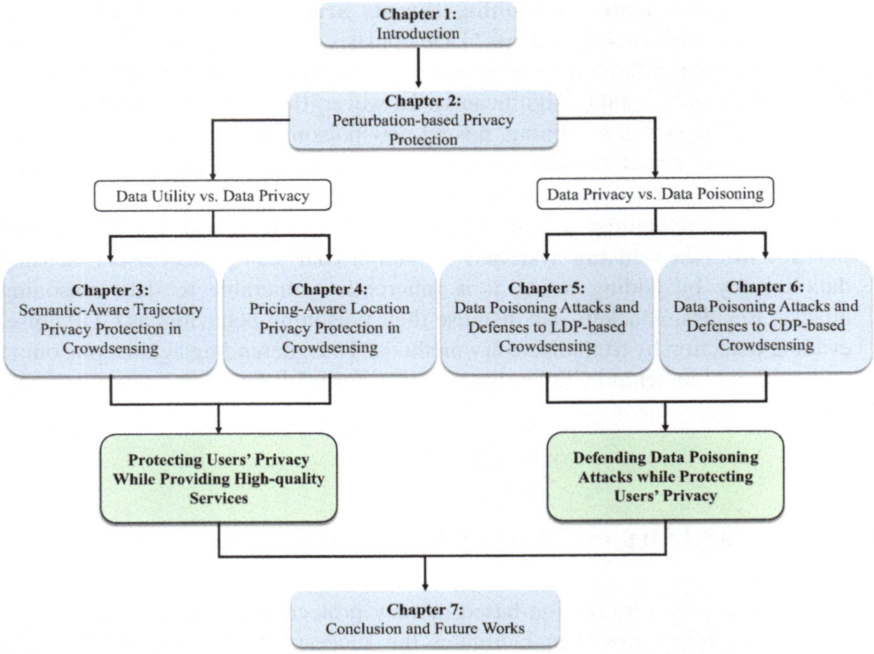

Fig. 1.4 The structure of this monograph

- **To defend against poisoning threats caused by differential privacy in crowdsensing systems (i.e., data privacy vs. data poisoning),** we propose data poisoning attacks and defenses to DP-based (including LDP and CDP) crowdsensing systems. By those two works, we have confirmed that DP can facilitate the success of data poisoning attacks against crowdsensing systems, and then proposed defense strategies to mitigate the poisoning threats.

1.3.2 Organization of This Monograph

As shown in Fig. 1.4, this monograph has 7 chapters. In Chap. 1, we introduce the background of crowdsensing systems and discuss privacy challenges from three dimensions: data utility, data privacy, and data poisoning. This is because the added noise perturbation inevitably damages data utility and facilitates the success of data poisoning attacks. Based on these challenges, we outline the aims and organization of this monograph.

In Chap. 2, we introduce three classic perturbation-based privacy notions: differential privacy, identifiability, and mutual-information privacy. Then, we establish a comprehensive privacy notions framework by analyzing the relations among these three notions. Powered by the framework, we conduct four research topics on perturbation-based privacy in crowdsensing systems from three dimensions: data utility, data poisoning, and data privacy, which are presented in Chaps. 3, 4, 5, and 6, respectively.

In Chap. 3, we focus on the trade-off between data utility and data privacy (i.e., data utility vs. data privacy), and propose a semantic-aware information-theoretic privacy protection mechanism, called SEITP. Specifically, SEITP protects users' trajectory data privacy and semantic privacy while preserving semantic-aware data utility. In addition, we theoretically prove that SEITP can prevent trajectory data and semantic privacy breaches, and experimentally show that it outperforms existing perturbation-based location privacy protection mechanisms.

In Chap. 4, we focus on the trade-off between data utility and data privacy (i.e., data utility vs. data privacy), and propose a pricing-aware differentially private framework for ride-on-demand services, called PADP-RoD. In this framework, users can obtain high-quality ride-matching and dynamic pricing services while protecting their data privacy. Besides, privacy analysis demonstrates that the framework can prevent users' location data privacy leakage, and experiment evaluations show it can provide users with high-quality services.

In Chap. 5, we focus on the trade-off between data privacy and data poisoning (i.e., data privacy vs. data poisoning), and propose data poisoning attacks and defenses to LDP-based crowdsensing systems. Specifically, the attackers can conceal their attack behavior by exploiting the LDP noise perturbation, verifying that LDP can pose poisoning threats to crowdsensing. To defend against the attacks, we also propose corresponding countermeasures. Additionally, evaluation results show that LDP facilitates the success of data poisoning attacks, and the defenses can mitigate these poisoning threats.

In Chap. 6, we focus on the trade-off between data privacy and data poisoning (i.e., data privacy vs. data poisoning), and propose data poisoning attacks and defenses to CDP-based crowdsensing systems. Particularly, we formulate the defenders-attackers interaction as a zero-sum Stackelberg game, wherein the attackers can evade the current defense strategy by hiding their poisoning behavior in the CDP noise perturbation. By this game, we find an effective defense strategy that mitigates poisoning threats caused by CDP. Furthermore, extensive experiments show the effectiveness of the proposed game-based defense approach.

Finally, we conclude the monograph and discuss future topics in Chap. 7.

References

1. Bankmycell: How many people have smartphones in the world? (2024). https://www.bankmycell.com/blog/how-many-phones-are-in-the-world
2. Mordor Intelligence: Wearable technology market size & share analysis - growth trends & forecasts (2024–2029) (2024). https://www.mordorintelligence.com/industry-reports/wearable-technology-market

3. Research and Markets: Network infrastructure market report 2024: A global SPSSlashDollar60.5+ billion market in 2023 driven by growing use of bandwidth-intensive applications for robust network infrastructure - forecast to 2030 (2024). https://www.globenewswire.com/news-release/2024/11/04/2973749/28124/en/Network-Infrastructure-Market-Report-2024-A-Global-60-5-Billion-Market-in-2023-Driven-by-Growing-Use-of-Bandwidth-Intensive-Applications-for-Robust-Network-Infrastructure-Forecast-.html
4. Howe, J.: The rise of crowdsourcing (2006). https://www.wired.com/2006/06/crowds/
5. Ganti, R.K., Ye, F., Lei, H.: Mobile crowdsensing: current state and future challenges. IEEE Commun. Mag. **49**(11), 32–39 (2011)
6. Surowiecki, J.: The Wisdom of Crowds: Why the Many are Smarter than the Few and How Collective Wisdom Shapes Business, Economies, Societies and Nations. Knopf Doubleday Publisher (2005)
7. Guo, B., Wang, Z., Yu, Z., Wang, Y., Yen, N.Y., Huang, R., Zhou, X.: Mobile crowd sensing and computing: The review of an emerging human-powered sensing paradigm. ACM Comput. Surv. **48**(1), 1–31 (2015)
8. Guo, B., Yu, Z., Zhou, X., Zhang, D.: From participatory sensing to mobile crowd sensing. In: Proceedings of IEEE International Conference on Pervasive Computing and Communication Workshops, pp. 593–598 (2014)
9. Capponi, A., Fiandrino, C., Kantarci, B., Foschini, L., Kliazovich, D., Bouvry, P.: A survey on mobile crowdsensing systems: Challenges, solutions, and opportunities. IEEE Commun. Surv. Tutorials **21**(3), 2419–2465 (2019)
10. Wang, Z., Cao, Y., Jiang, K., Zhou, H., Kang, J., Zhuang, Y., Tian, D., Leung, V.C.M.: When crowdsensing meets smart cities: A comprehensive survey and new perspectives. IEEE Commun. Surv. Tutorials, 1–52 (2024). https://doi.org/10.1109/COMST.2024.3400121
11. Yu, S., Chen, X., Wang, S., Pu, L., Wu, D.: An edge computing-based photo crowdsourcing framework for real-time 3d reconstruction. IEEE Trans. Mob. Comput. **21**(2), 421–432 (2022)
12. de Baar, J.H.S., van der Schrier, G.: "come as you are": Reconsidering the need for complex quality control when gridding crowd-sourced weather data. Q. J. R. Meteorol. Soc. **151**(766), 1–17 (2024)
13. Hu, S., Su, L., Liu, H., Wang, H., Abdelzaher, T.F.: Smartroad: Smartphone-based crowd sensing for traffic regulator detection and identification. ACM Trans. Netw. **11**(4), 1–27 (2015)
14. Asiminidis, C., Kokkonis, G., Kontogiannis, S.: Ble sniffing for crowd sensing and directionality scanning of mobile devices inside tunnels. In: Proceedings of World Symposium on Communication Engineering, pp. 54–58 (2020)
15. Davari, M., Amintoosi, H.: A survey on participant recruitment in crowdsensing systems. In: Proceedings of International Conference on Computer and Knowledge Engineering, pp. 286–291 (2016)
16. Shang, L., Zhang, Y., Ye, Q., Speir, S.L., Peters, B.W., Wu, Y., Stoffel, C.J., Bolster, D., Tank, J.L., Wood, D.M., Wei, N., Wang, D.: Crowdwatersens: An uncertainty-aware crowdsensing approach to groundwater contamination estimation. Pervasive Mob. Comput. **92**, 1–18 (2023)
17. Zappatore, M., Longo, A., Bochicchio, M.A.: Using mobile crowd sensing for noise monitoring in smart cities. In: Proceedings of International Multidisciplinary Conference on Computer and Energy Science, pp. 1–6 (2016)
18. Costa, C., Laoudias, C., Zeinalipour-Yazti, D., Gunopulos, D.: Smarttrace: Finding similar trajectories in smartphone networks without disclosing the traces. In: Proceedings of IEEE International Conference on Data Engineering, pp. 1288–1291 (2011)
19. Zheng, Y., Liu, F., Hsieh, H.P.: U-air: when urban air quality inference meets big data. In: Proceedings of ACM SIGKDD International Conference on Knowledge Discovery and Data Mining, pp. 1436–1444 (2013)
20. Maisonneuve, N., Stevens, M., Ochab, B.: Participatory noise pollution monitoring using mobile phones. Inf. Polity **15**, 51–71 (2010)
21. Karamshuk, D., Boldrini, C., Conti, M., Passarella, A.: Human mobility models for opportunistic networks. IEEE Commun. Mag. **49**(12), 157–165 (2011)
22. Tang, L., Liu, H.: Community Detection and Mining in Social Media. Springer Nature (2022)

23. Du, R., Yu, Z., Mei, T., Wang, Z., Wang, Z., Guo, B.: Predicting activity attendance in event-based social networks: content, context and social influence. In: Proceedings of ACM International Joint Conference on Pervasive and Ubiquitous Computing, pp. 425–434 (2014)
24. Yang, D., Zhang, D., Yu, Z., Yu, Z.: Fine-grained preference-aware location search leveraging crowdsourced digital footprints from LBSNs. In: Proceedings of ACM International Joint Conference on Pervasive and Ubiquitous Computing, pp. 479–488 (2013)
25. Yu, Z., Feng, Y., Xu, H., Zhou, X.: Recommending travel packages based on mobile crowdsourced data. IEEE Commun. Mag. **52**(8), 56–62 (2014)
26. Ballesteros, J., Carbunar, B., Rahman, M., Rishe, N., Iyengar, S.: Towards safe cities: A mobile and social networking approach. IEEE Trans. Parallel Distrib. Syst. **25**(9), 2451–2462 (2014)
27. Sakaki, T., Okazaki, M., Matsuo, Y.: Earthquake shakes twitter users: real-time event detection by social sensors. In: Proceedings of International Conference on World Wide Web, pp. 851–860 (2010)
28. Eriksson, J., Girod, L., Hull, B., Newton, R., Madden, S., Balakrishnan, H.: The pothole patrol: using a mobile sensor network for road surface monitoring. In: Proceedings of International Conference on Mobile Systems, Applications, and Services, pp. 29–39 (2008)
29. Ferguson, N.M., Cummings, D.A.T., Fraser, C., Cajka, J.C., Cooley, P.C., Burke, D.S.: Strategies for mitigating an influenza pandemic. Nature **7010**(10), 448–452 (2006)
30. Reddy, S., Parker, A., Hyman, J., Burke, J., Estrin, D., Hansen, M.: Image browsing, processing, and clustering for participatory sensing: lessons from a dietsense prototype. In: Proceedings of Workshop on Embedded Networked Sensors, pp. 13–17 (2007)
31. ISACA: Privacy in practice 2024 report (2024). https://www.isaca.org/resources/reports/privacy-in-practice-2024-report?utm_source=chatgpt.com#LeadForm
32. Reuters: EU privacy regulator fines meta 91 million euros over password storage (2024). https://www.reuters.com/technology/eu-privacy-regulator-fines-meta-91-million-euros-over-password-storage-2024-09-27/
33. Reuters: Chinese electric automaker Nio hit by data breach (2022). https://www.reuters.com/legal/government/chinese-electric-automaker-nio-hit-by-data-breach-2022-12-20/
34. Li, Z., Zheng, Z., Guo, S., Guo, B., Xiao, F., Ren, K.: Disguised as privacy: Data poisoning attacks against differentially private crowdsensing systems. IEEE Trans. Mob. Comput. **22**(9), 5155–5169 (2023)
35. Zheng, Z., Li, Z., Huang, C., Long, S., Li, M., Shen, X.: Data poisoning attacks and defenses to LDP-based privacy-preserving crowdsensing. IEEE Trans. Dependable Secur. Comput. **21**(5), 4861–4878 (2024)
36. Zheng, Z., Li, Z., Huang, C., Long, S., Shen, X.: Defending data poisoning attacks in DP-based crowdsensing: A game-theoretic approach. IEEE Trans. Mob. Comput. **24**(3), 1859–1876 (2025)
37. Cheng, X., He, B., Li, G., Cheng, B.: A survey of crowdsensing and privacy protection in digital city. IEEE Trans. Comput. Soc. Syst. **10**(6), 3471–3487 (2023)
38. Zheng, Z., Li, Z., Li, J., Jiang, H., Li, T., Guo, B.: Utility-aware and privacy-preserving trajectory synthesis model that resists social relationship privacy attacks. ACM Trans. Intell. Syst. Technol. **13**(3), 1–28 (2022)
39. Zheng, Z., Li, Z., Jiang, H., Zhang, L.Y., Tu, D.: Semantic-aware privacy-preserving online location trajectory data sharing. IEEE Trans. Inf. Forensics Secur. **17**, 2256–2271 (2022)
40. De Montjoye, Y.A., Hidalgo, C.A., Verleysen, M., Blondel, V.D.: Unique in the crowd: The privacy bounds of human mobility. Sci. Rep. **3**(1), 1–5 (2013)
41. Gambs, S., Killijian, M.O., del Prado Cortez, M.N.: Show me how you move and I will tell you who you are. In: Proceedings of ACM SIGSPATIAL International Workshop on Security and Privacy in GIS and LBS, pp. 34–41 (2010)
42. Thomas, K., McCoy, D., Grier, C., Kolcz, A., Paxson, V.: Trafficking fraudulent accounts: The role of the underground market in twitter spam and abuse. In: Proceedings of USENIX Security Symposium, pp. 195–210 (2013)

43. Giraldo, J., Cardenas, A., Kantarcioglu, M., Katz, J.: Adversarial classification under differential privacy. In: Proceedings of Network and Distributed Systems Security Symposium, pp. 1–18 (2020)
44. Cao, X., Jia, J., Gong, N.Z.: Data poisoning attacks to local differential privacy protocols. In: Proceedings of USENIX Security Symposium, pp. 947–964 (2021)
45. Andrés, M.E., Bordenabe, N.E., Chatzikokolakis, K., Palamidessi, C.: Geo-indistinguishability: Differential privacy for location-based systems. In: Proceedings of ACM SIGSAC Conference on Computer and Communications Security, pp. 901–914 (2013)
46. Sun, M., Tay, W.P.: On the relationship between inference and data privacy in decentralized IoT networks. IEEE Trans. Inf. Forensics Secur. **15**, 852–866 (2020)
47. Jiang, H., Zhao, P., Wang, C.: Roblop: Towards robust privacy preserving against location dependent attacks in continuous LBS queries. IEEE/ACM Trans. Netw. **26**(2), 1018–1032 (2018)
48. Wang, H., Xu, Z., Jia, S., Xia, Y., Zhang, X.: Why current differential privacy schemes are inapplicable for correlated data publishing? World Wide Web **24**(1), 1573–1413 (2021)
49. Li, Q., Gundersen, J.S., Heusdens, R., Christensen, M.G.: Privacy-preserving distributed processing: Metrics, bounds and algorithms. IEEE Trans. Inf. Forensics Secur. **16**, 2090–2103 (2021)
50. Wang, C.X., Song, Y., Tay, W.P.: Arbitrarily strong utility-privacy tradeoff in multi-agent systems. IEEE Trans. Inf. Forensics Secur. **16**, 671–684 (2021)
51. Zhang, W., Li, M., Tandon, R., Li, H.: Online location trace privacy: An information theoretic approach. IEEE Trans. Inf. Forensics Secur. **14**(1), 235–250 (2019)
52. Guo, S., Chen, C., Wang, J., Liu, Y., Xu, K., Zhang, D., Chiu, D.M.: A simple but quantifiable approach to dynamic price prediction in ride-on-demand services leveraging multi-source urban data. In: Proceedings of the ACM Interactive, Mobile, Wearable and Ubiquitous Technologies, vol. 2(3), pp. 1–24 (2018)
53. Guo, S., Chen, C., Liu, Y., Xu, K., Chiu, D.M.: Modelling passengers' reaction to dynamic prices in ride-on-demand services: A search for the best fare. In: Proceedings of the ACM Interactive, Mobile, Wearable and Ubiquitous Technologies, vol. 1(4), pp. 1–23 (2018)
54. Guo, S., Chen, C., Wang, J., Liu, Y., Xu, K., Yu, Z., Zhang, D., Chiu, D.M.: Rod-revenue: Seeking strategies analysis and revenue prediction in ride-on-demand service using multi-source urban data. IEEE Trans. Mob. Comput. **19**(9), 2202–2220 (2020)
55. Guo, S., Chen, C., Wang, J., Ding, Y., Liu, Y., Xu, K., Yu, Z., Zhang, D.: A force-directed approach to seeking route recommendation in ride-on-demand service using multi-source urban data. IEEE Trans. Mob. Comput. **21**(6), 1909–1926 (2022)
56. Chen, M.K.: Dynamic pricing in a labor market: Surge pricing and flexible work on the Uber platform. In: Proceedings of ACM Conference on Economics and Computation, pp. 1–19 (2016)
57. Huang, J., Huang, L., Liu, M., Li, H., Tan, Q., Ma, X., Cui, J., Huang, D.S.: Deep reinforcement learning-based trajectory pricing on ride-hailing platforms. ACM Trans. Intell. Syst. Technol. **13**(3), 1–19 (2022)
58. Shao, S., Mittal, A., Twumasi-Boakye, R., Gupta, A.: Fair pricing of ridehailing services with asymmetric demand and travel time. IEEE Trans. Control Netw. Syst. **9**(2), 670–681 (2022)
59. Wang, L., Yang, D., Han, X., Wang, T., Zhang, D., Ma, X.: Location privacy-preserving task allocation for mobile crowdsensing with differential geo-obfuscation. In: Proceedings of International Conference on World Wide Web, pp. 627–636 (2017)
60. Wang, H., Wang, E., Yang, Y., Wu, J., Dressler, F.: Privacy-preserving online task assignment in spatial crowdsourcing: A graph-based approach. In: Proceedings of IEEE International Conference on Computer Communications, pp. 570–579 (2022)
61. Tao, Q., Tong, Y., Zhou, Z., Shi, Y., Chen, L., Xu, K.: Differentially private online task assignment in spatial crowdsourcing: A tree-based approach. In: Proceedings of IEEE International Conference on Data Engineering, pp. 517–528 (2020)

62. To, H., Shahabi, C., Xiong, L.: Privacy-preserving online task assignment in spatial crowdsourcing with untrusted server. In: Proceedings of IEEE International Conference on Data Engineering, pp. 833–844 (2018)
63. Wang, L., Zhang, D., Yang, D., Lim, B.Y., Han, X., Ma, X.: Sparse mobile crowdsensing with differential and distortion location privacy. IEEE Trans. Inf. Forensics Secur. **15**, 2735–2749 (2020)
64. Wang, L., Yang, D., Han, X., Zhang, D., Ma, X.: Mobile crowdsourcing task allocation with differential-and-distortion geo-obfuscation. IEEE Trans. Dependable Secur. Comput. **18**(2), 967–981 (2021)
65. Fang, M., Sun, M., Li, Q., Gong, N.Z., Tian, J., Liu, J.: Data poisoning attacks and defenses to crowdsourcing systems. In: Proceedings of International World Wide Web Conference, pp. 969–980 (2021)
66. Miao, C., Li, Q., Su, L., Huai, M., Jiang, W., Gao, J.: Attack under disguise: An intelligent data poisoning attack mechanism in crowdsourcing. In: Proceedings of International World Wide Web Conference, pp. 13–22 (2018)
67. Miao, C., Li, Q., Xiao, H., Jiang, W., Huai, M., Su, L.: Towards data poisoning attacks in crowd sensing systems. In: Proceedings of ACM International Symposium on Mobile Ad Hoc Networking and Computing, pp. 111–120 (2018)
68. Zhao, Y., Gong, X., Lin, F., Chen, X.: Data poisoning attacks and defenses in dynamic crowdsourcing with online data quality learning. IEEE Trans. Mob. Comput. **22**(5), 2569–2581 (2023)
69. Zhang, H., Li, M.: Multi-round data poisoning attack and defense against truth discovery in crowdsensing systems. In: Proceedings of IEEE International Conference on Mobile Data Management, pp. 109–118 (2022)
70. Fujimoto, R., Kamiyama, N.: Poisoning attacks in crowdsensing over multiple areas. In: Proceedings of IEEE Global Communications Conference, pp. 68–73 (2022)
71. Huang, Z., Pan, M., Gong, Y.: Robust truth discovery against data poisoning in mobile crowdsensing. In: Proceedings of IEEE Global Communications Conference, pp. 1–6 (2019)
72. Zhang, H., Li, M., Sun, Y., Qu, G.: Robust truth discovery against multi-round data poisoning attacks. In: Proceedings of Wireless Algorithms, Systems, and Applications, pp. 258–270 (2022)
73. Cheu, A., Smith, A., Ullman, J.: Manipulation attacks in local differential privacy. In: Proceedings of IEEE Symposium on Security and Privacy, pp. 883–900 (2021)
74. Wu, Y., Cao, X., Jia, J., Gong, N.Z.: Poisoning attacks to local differential privacy protocols for Key-Value data. In: Proceedings of USENIX Security Symposium, pp. 519–536 (2022)
75. Li, X., Gong, N.Z., Li, N., Sun, W., Li, H.: Fine-grained poisoning attacks to local differential privacy protocols for mean and variance estimation. In: Proceedings of USENIX Security Symposium, pp. 1739–1756 (2023)
76. Cheu, A., Zhilyaev, M.: Differentially private histograms in the shuffle model from fake users. In: Proceedings of IEEE Symposium on Security and Privacy, pp. 440–457 (2022)
77. Imola, J., Chowdhury, A.R., Chaudhuri, K.: Robustness of locally differentially private graph analysis against poisoning. Preprint. arXiv:2210.14376 (2022)

Chapter 2
Perturbation-Based Privacy Preservation

2.1 Classical Privacy Notions

Consider a sensing dataset x consisting of n values drawn from a finite domain $\mathcal{D}$, denoted as $x = \{d_1, \cdots, d_n\}$, where each value represents a personal sensing data. Let $\mathcal{D}^n$ denote the set of all possible dataset x, $\forall x \in \mathcal{D}^n$. To protect users' privacy, as shown in Fig. 2.1, we add appropriate noise to the personal data before sharing [1–6]. Specifically, in the central scenario [1, 7–9], users upload their real sensing data to a trusted server, which then perturbs the observed sensing dataset x into the noisy dataset y (denoted as $y = \{e_1, \cdots, e_n\}, \forall y \in \mathcal{D}^n$) by performing a perturbation-based privacy-preserving mechanism $\mathcal{A}$. In the local scenario [2, 10–12], users locally perturb their real sensing data $d, \forall d \in \mathcal{D}$ into the noisy data $e, \forall e \in \mathcal{D}$ utilizing the mechanism $\mathcal{A}$. Next, we introduce the definition of neighboring datasets [13, 14], which is a fundamental concept in central privacy notions.

Definition 2.1 (Neighboring Datasets) Let a dataset x consisting of n values drawn from a finite domain $\mathcal{D}$. Then, two datasets x and x' are neighbors, denoted by $x \sim x', \forall x, x' \in \mathcal{D}^n$, if and only if they differ in exactly one value.

Powered by Definition 2.1, we briefly overview differential privacy (DP) [13–16], identifiability [17–19], and mutual-information privacy [18, 20–24] from central and local scenarios.

2.1.1 Identifiability

The notion of identifiability [17, 18] is based on the indistinguishability between any two datasets from the Bayesian view. Then, we give detailed definitions of identifiability in both central and local scenarios below.

Z. Zheng et al., *Perturbation Based Privacy in Crowdsensing*, Wireless Networks,
https://doi.org/10.1007/978-3-031-95052-0_2

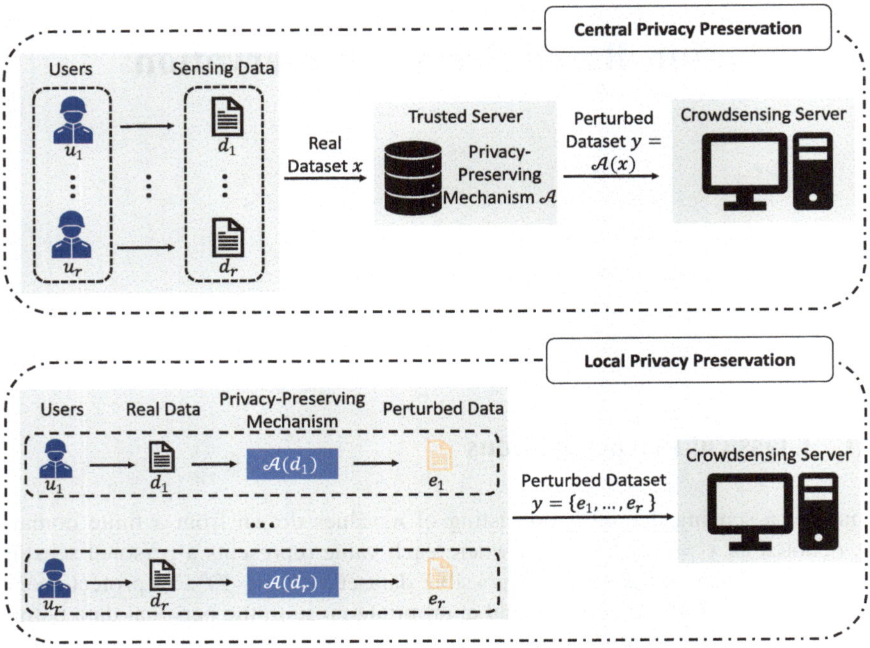

Fig. 2.1 The workflow of perturbation-based privacy-preserving mechanisms

Definition 2.2 (Central Identifiability) A central mechanism $\mathcal{A} := p_{Y|X}(y \mid x)$ satisfies ε_i-identifiability for some $\varepsilon_i \in \mathbb{R}^+$ if for any pair of neighboring datasets $x, x' \in \mathcal{D}^n$ and any $y \in \mathcal{D}^n$,

$$p_{X|Y}(x \mid y) \leq e^{\varepsilon_i} \cdot p_{X|Y}(x' \mid y). \tag{2.1}$$

Definition 2.3 (Local Identifiability) A local mechanism $\mathcal{A} := p_{E|D}(e \mid d)$ satisfies ε_i-identifiability for some $\varepsilon_i \in \mathbb{R}^+$ if for any two real data $d_i, d_j \in \mathcal{D}$ and any $e \in \mathcal{D}$,

$$p_{D|E}(d_i \mid e) \leq e^{\varepsilon_i} \cdot p_{D|E}(d_j \mid e). \tag{2.2}$$

The definition of central identifiability elucidates a crucial characteristic: attackers are unable to discern differentiation between any two neighboring datasets, i.e., $\forall x, x' \in \mathcal{D}^n : x \sim x'$, based on their posterior probabilities given the perturbed dataset y. Similarly, according to the definition of local identifiability, attackers cannot distinguish between any two real data, i.e., $\forall d_i, d_j \in \mathcal{D}$, using their posterior probabilities given the observed data e. As a result, the real data of any individual is difficult for attackers to identify when the mechanism $\mathcal{A}$ satisfies ε_i-identifiability with a small ε_i.

2.1.2 Differential Privacy

DP [13, 15], unlike identifiability, provides a relative guarantee regarding disclosures. Specifically, DP ensures that the probability of privacy breaches will increase by at most a multiplicative factor when the perturbed dataset (or data) is published. The detailed definitions of DP in both central and local scenarios are given below.

Definition 2.4 (Central Differential Privacy) A mechanism $\mathcal{A} := p_{Y|X}(y \mid x)$ satisfies ε_d-central differential privacy (ε_d-CDP) for some $\varepsilon_d \in \mathbb{R}^+$ if for any pair of neighboring datasets $x, x' \in \mathcal{D}^n$ and any $y \in \mathcal{D}^n$,

$$p_{Y|X}(y \mid x) \leq e^{\varepsilon_d} \cdot p_{Y|X}(y \mid x'). \tag{2.3}$$

Definition 2.5 (Local Differential Privacy) A mechanism $\mathcal{A} := p_{E|D}(e \mid d)$ satisfies ε_d-local differential privacy (ε_d-LDP) for some $\varepsilon_d \in \mathbb{R}^+$ if for any real data $d_i, d_j \in \mathcal{D}$ and any perturbed data $e \in \mathcal{D}$,

$$p_{E|D}(e \mid d_i) \leq e^{\varepsilon_d} \cdot p_{E|D}(e \mid d_j). \tag{2.4}$$

According to the above definitions, we can observe that DP ensures the indistinguishability of conditional probabilities (i.e., $p_{X|Y}$ or $p_{D|E}$) for a small privacy budget ε_d. As a result, even if attackers are aware of the privacy-preserving mechanism $\mathcal{A}$, they cannot identify the real data when the mechanism $\mathcal{A}$ satisfies ε_d-DP with a small ε_d.

2.1.3 Mutual-Information Privacy

Mutual information [25] is an information-theoretic concept that quantifies the average amount of information contained in one random variable about another. For example, the mutual information between random variables X and Y, i.e., $I(X; Y)$, measures the privacy leakage of the real dataset x arising from sharing the perturbed dataset y. Specifically, the mutual information $I(X; Y)$ is minimized to 0 when X and Y are independent, and maximized to entropy $H(X)$ when Y equals X. Thus, smaller mutual information implies stronger privacy preservation. Then, we provide detailed definitions of mutual-information privacy in both central and local scenarios below.

Definition 2.6 (Central Mutual-Information Privacy) A central mechanism $\mathcal{A} := p_{Y|X}(y \mid x)$ satisfies ε_m-mutual-information privacy for some $\varepsilon_m \in \mathbb{R}^+$ if the mutual information between random variables X and Y satisfies $I(X; Y) \leq \varepsilon_m$, where

$$I(X; Y) = \sum_{x, y \in \mathcal{D}^n} p_{X,Y}(x, y) \cdot \log \frac{p_{X,Y}(x, y)}{p_X(x) p_Y(y)}. \tag{2.5}$$

Definition 2.7 (Local Mutual-Information Privacy) A local mechanism $\mathcal{A} := p_{E|D}(e \mid d)$ satisfies ε_m-mutual-information privacy for some $\varepsilon_m \in \mathbb{R}^+$ if the mutual information between random variables D and E satisfies $I(D; E) \leq \varepsilon_m$, where

$$I(D; E) = \sum_{d,e \in \mathcal{D}} p_{D,E}(d, e) \cdot \log \frac{p_{D,E}(d, e)}{p_D(d) p_E(e)}. \tag{2.6}$$

2.2 Relations Between Privacy Notions

Before outlining the relations between identifiability, DP, and mutual-information privacy, we first define the minimum prior probability difference σ_{min} (i.e., Definition 2.8), and then introduce its equivalent forms (i.e., Lemma 2.1).

Definition 2.8 (Minimum Prior Probability Difference) Let p_X and p_D be the prior probability of X (a random variable to real dataset x) and D (a random variable to real data d), respectively. In the central scenario, the minimum prior probability difference between neighboring datasets is defined as

$$\sigma^c_{min} = \min_{x,x' \in \mathcal{D}^n : x \sim x'} \log \frac{p_X(x)}{p_X(x')}. \tag{2.7}$$

In the local scenario, the minimum prior probability difference between any two data is defined as

$$\sigma^l_{min} = \min_{d_i, d_j \in \mathcal{D}} \log \frac{p_D(d_i)}{p_D(d_j)}, \tag{2.8}$$

Moreover, we refer to σ^c_{min} and σ^l_{min} collectively as the minimum prior probability difference, denoted as σ_{min}, when distinguishing between central and local scenarios is unnecessary.

Lemma 2.1 *The minimum prior probability difference has the following equivalent formulation:*

$$\sigma^c_{min} = \min_{x,x' \in \mathcal{D}^n : x \sim x'} \log \frac{p_X(x')}{p_X(x)}, \tag{2.9}$$

$$\sigma^l_{min} = \min_{d_j, d_i \in \mathcal{D}} \log \frac{p_D(d_j)}{p_D(d_i)}. \tag{2.10}$$

Specifically, $\sigma^c_{min} = 0$ *(or* $\sigma^l_{min} = 0$*) if and only if the random variable* X *(or* D*) follows a uniform distribution.*

Proof The symbols $x \sim x', \forall x, x' \in \mathcal{D}^n$ and $x' \sim x, \forall x, x' \in \mathcal{D}^n$ are equivalent, both representing that x and x' are neighboring datasets. As a result, we have

$$\min_{x,x' \in \mathcal{D}^n : x \sim x'} \log \frac{p_X(x')}{p_X(x)} = \min_{x,x' \in \mathcal{D}^n : x' \sim x} \log \frac{p_X(x)}{p_X(x')}, \tag{2.11}$$

that is, Eqs. (2.7) and (2.9) are equivalent. Similarly, Eqs. (2.8) and (2.10) are equivalent. Specifically, if the random variable X (or D) follows a uniform distribution, i.e., $p_X(x) = p_X(x'), \forall x, x' \in \mathcal{D}^n$ (or $p_D(d_i) = p_D(d_j), \forall d_i, d_j \in \mathcal{D}$), we have $\sigma^c_{min} = 0$ (or $\sigma^l_{min} = 0$). □

Powered by Definition 2.8 and Lemma 2.1, we propose a completeness privacy notions framework, as shown in Fig. 2.2, which involves three main conclusions. First, the privacy-preserving mechanism $\mathcal{A}$ that satisfies ε_i-identifiability also satisfies ε_d-DP with $\varepsilon_d = \varepsilon_i - \sigma_{min}$ and ε_m-mutual-information privacy with $\varepsilon_m = 2(\varepsilon_i - \sigma_{min})$. Second, the mechanism $\mathcal{A}$ that satisfies ε_d-DP also satisfies ε_i-identifiability with $\varepsilon_i = \varepsilon_d + \sigma_{min}$ and ε_m-mutual-information privacy with $\varepsilon_m = 2\varepsilon_d$. Third, the mechanism $\mathcal{A}$ that satisfies ε_m-mutual-information privacy does not guarantee ε_i-identifiability and ε_d-DP, i.e., the values of ε_i and ε_d may approach infinity when ε_m is a bounded constant. Obviously, we can consistently obtain the relations between ε_i-identifiability, ε_d-DP, and ε_m-mutual-information privacy from the proposed framework, referred to as completeness. For example, according to ε_i-identifiability implies $(\varepsilon_i - \sigma_{min})$-DP and ε_d-DP implies $(\varepsilon_d + \sigma_{min})$-identifiability, we can deduce that the mechanism $\mathcal{A}$ satisfying ε_i-identifiability also satisfies $(\varepsilon_i - \sigma_{min} + \sigma_{min})$-identifiability (i.e., this phenomenon is referred to as completeness). Next, we provide detailed proofs of the proposed privacy framework.

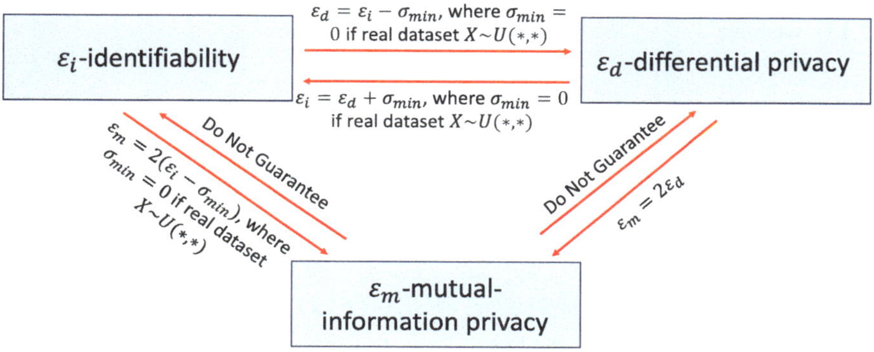

Fig. 2.2 A completeness privacy notions framework: relations between identifiability, differential privacy, and mutual-information privacy

2.2.1 Identifiability vs. Differential Privacy

Identifiability and DP define the indistinguishability of any neighboring datasets in the finite domain $\mathcal{D}^n$ (or any data in the finite domain $\mathcal{D}$) based on the posterior probability $p_{X|Y}$ (or $p_{D|E}$) and the conditional probability $p_{Y|X}$ (or $p_{E|D}$), respectively. According to definitions of identifiability, DP, and minimum prior probability difference, we prove that the mechanism $\mathcal{A}$ satisfying ε_i-identifiability also satisfies ε_d-DP with $\varepsilon_d = \varepsilon_i - \sigma_{min}$, and the mechanism $\mathcal{A}$ satisfying ε_d-DP also satisfies ε_i-identifiability with $\varepsilon_i = \varepsilon_d + \sigma_{min}$. The detailed theorems and corresponding proofs are given below.

Theorem 2.1 **(ε_i-Identifiability Implies ($\varepsilon_i - \sigma_{min}$)-Differential Privacy)** *Given the minimum prior probability difference σ_{min} with $\sigma_{min} \geq 0$. The mechanism $\mathcal{A}$ also satisfies ε_d-DP with $\varepsilon_d = \varepsilon_i - \sigma_{min}$ when it satisfies ε_i-identifiability.*

Proof We first prove that ε_i-identifiability implies ($\varepsilon_i - \sigma_{min}$)-DP in the central scenario, and then prove the same in the local scenario. According to Definition 2.8, i.e., the minimum prior probability difference between neighboring datasets in the central scenario, we have

$$\frac{1}{p_X(x)} \leq e^{-\sigma_{min}^c} \cdot \frac{1}{p_X(x')}, \ \forall x, x' \in \mathcal{D}^n : x \sim x'. \tag{2.12}$$

Additionally, since the mechanism $\mathcal{A} := p_{Y|X}(y \mid x)$ satisfies ε_i-central identifiability (detailed in Definition 2.2), we have

$$p_{X|Y}(x \mid y) \leq e^{\varepsilon_i} \cdot p_{X|Y}(x' \mid y), \ \forall x, x' \in \mathcal{D}^n : x \sim x'. \tag{2.13}$$

For any neighboring datasets, i.e., $\forall x, x' \in \mathcal{D}^n : x \sim x'$, powered by Eqs. (2.12)–(2.13), we have

$$p_{X|Y}(x \mid y) \cdot \frac{1}{p_X(x)} \leq e^{\varepsilon_i} \cdot p_{X|Y}(x' \mid y) \cdot \frac{1}{p_X(x)} \tag{2.14}$$

$$\leq e^{\varepsilon_i - \sigma_{min}^c} \cdot p_{X|Y}(x' \mid y) \cdot \frac{1}{p_X(x')}. \tag{2.15}$$

Since

$$\frac{p_{X|Y}(x \mid y)}{p_X(x)} = \frac{p_{Y|X}(y \mid x)}{p_Y(y)}, \tag{2.16}$$

$$\frac{p_{X|Y}(x' \mid y)}{p_X(x')} = \frac{p_{Y|X}(y \mid x')}{p_Y(y)}, \tag{2.17}$$

Eq. (2.15) can be reformulated as

$$\frac{p_{Y|X}(y \mid x)}{p_Y(y)} \leq e^{\varepsilon_i - \sigma^c_{min}} \cdot \frac{p_{Y|X}(y \mid x')}{p_Y(y)}, \ \forall x, x' \in \mathcal{D}^n : x \sim x'. \tag{2.18}$$

Therefore, as detailed in Definition 2.4, the mechanism $\mathcal{A} := p_{Y|X}(y \mid x)$ satisfies $(\varepsilon_i - \sigma^c_{min})$-CDP. Similarly, we can prove that the mechanism $\mathcal{A} := p_{E|D}(e \mid d)$ satisfying ε_i-local identifiability also satisfies ε_d-LDP with $\varepsilon_d = \varepsilon_i - \sigma^l_{min}$. In summary, ε_i-identifiability implies $(\varepsilon_i - \sigma_{min})$-DP. □

Theorem 2.2 (ε_d-Differential Privacy Implies $(\varepsilon_d + \sigma_{min})$-Identifiability) *Given the minimum prior probability difference σ_{min} with $\sigma_{min} \geq 0$. The mechanism $\mathcal{A}$ also satisfies ε_i-identifiability with $\varepsilon_i = \varepsilon_d + \sigma_{min}$ when it satisfies ε_d-DP.*

Proof We first prove that ε_d-CDP implies $(\varepsilon_d + \sigma^c_{min})$-central identifiability, and then prove that ε_d-LDP implies $(\varepsilon_d + \sigma^l_{min})$-local identifiability. According to Lemma 2.1, i.e., the equivalent formulation of minimum prior probability difference between neighboring datasets, we have

$$p_X(x) \leq e^{\sigma^c_{min}} \cdot p_X(x'), \ \forall x, x' \in \mathcal{D}^n : x \sim x'. \tag{2.19}$$

In addition, we have

$$p_{Y|X}(y \mid x) \leq e^{\varepsilon_d} \cdot p_{Y|X}(y \mid x'), \ \forall x, x' \in \mathcal{D}^n : x \sim x', \tag{2.20}$$

due to the mechanism $\mathcal{A} := p_{Y|X}(y \mid x)$ satisfies ε_d-CDP. Powered by Eqs. (2.19)–(2.20), for any neighboring datasets, i.e., $\forall x, x' \in \mathcal{D}^n : x \sim x'$, we have

$$p_{Y|X}(y \mid x) \cdot p_X(x) \leq e^{\varepsilon_d} \cdot p_{Y|X}(y \mid x') \cdot p_X(x) \tag{2.21}$$

$$\leq e^{\varepsilon_d + \sigma^c_{max}} \cdot p_{Y|X}(y \mid x') \cdot p_X(x'). \tag{2.22}$$

Equation (2.22) can be reformulated as

$$p_{X|Y}(x \mid y) \cdot p_Y(y) \leq e^{\varepsilon_d + \sigma^c_{max}} \cdot p_{X|Y}(x' \mid y) \cdot p_Y(y), \ \forall x, x' \in \mathcal{D}^n : x \sim x', \tag{2.23}$$

since

$$p_{Y|X}(y \mid x) \cdot p_X(x) = p_{X|Y}(x \mid y) \cdot p_Y(y), \tag{2.24}$$

$$p_{Y|X}(y \mid x') \cdot p_X(x') = p_{X|Y}(x' \mid y) \cdot p_Y(y), \tag{2.25}$$

According to Eq. (2.23), we deduce that the mechanism $\mathcal{A} := p_{Y|X}(y \mid x)$ satisfies $(\varepsilon_d + \sigma^c_{min})$-central identifiability (detailed in Definition 2.2). Similarly, we can prove that the mechanism $\mathcal{A} := p_{E|D}(e \mid d)$ satisfying ε_d-LDP also satisfies ε_i-local identifiability with $\varepsilon_i = \varepsilon_d + \sigma^l_{min}$. In summary, ε_d-DP implies $(\varepsilon_d + \sigma_{min})$-identifiability. □

2.2.2 Identifiability vs. Mutual-Information Privacy

Identifiability quantifies the indistinguishability of any neighboring dataset in the finite domain $\mathcal{D}^n$ (or any data in the finite domain $\mathcal{D}$) from the Bayesian view, while mutual-information privacy measures the average privacy leakage about the real dataset x (or real data d) caused by sharing perturbed dataset y (or perturbed data e). According to definitions of identifiability, mutual-information privacy, and minimum prior probability difference, we prove that the mechanism $\mathcal{A}$ satisfying ε_i-identifiability also satisfies ε_m-mutual-information privacy with $\varepsilon_m = 2(\varepsilon_i - \sigma_{min})$, and the mechanism $\mathcal{A}$ satisfying ε_m-mutual-information privacy does not guarantee ε_i-identifiability. Then, we give detailed theorems and corresponding proofs below.

Theorem 2.3 (ε_i-Identifiability Implies $2(\varepsilon_i - \sigma_{min})$-Mutual-Information Privacy) *The mechanism $\mathcal{A}$ also satisfies ε_m-mutual-information privacy with $\varepsilon_m = 2(\varepsilon_i - \sigma_{min})$ when it satisfies ε_i-identifiability.*

Proof In the central scenario, since the mechanism $\mathcal{A} := p_{Y|X}(y \mid x)$ satisfies ε_i-identifiability, as defined in Definition 2.2, we have

$$\frac{p_{X|Y}(x \mid y)}{p_{X|Y}(x' \mid y)} \le e^{\varepsilon_i}, \ \forall x, x' \in \mathcal{D}^n : x \sim x'. \tag{2.26}$$

Taking the logarithm of both sides of Eq. (2.26) gives the following

$$\log \frac{p_{X|Y}(x \mid y)}{p_{X|Y}(x' \mid y)} \le \varepsilon_i, \ \forall x, x' \in \mathcal{D}^n : x \sim x'. \tag{2.27}$$

Therefore, we have

$$\log \frac{p_{X,Y}(x, y)}{p_X(x) p_Y(y)} \tag{2.28}$$

$$\overset{(a)}{\le} \varepsilon_i - \log \frac{p_X(x)}{p_X(x')} - \log \frac{\sum_{x \in \mathcal{D}^n} p_{Y|X}(y \mid x) p_X(x)}{p_{Y|X}(y \mid x')} \tag{2.29}$$

$$\overset{(b)}{\le} \varepsilon_i - \sigma^c_{min} - \log \frac{\sum_{x \in \mathcal{D}^n} p_{Y|X}(y \mid x) p_X(x)}{p_{Y|X}(y \mid x')} \tag{2.30}$$

$$\overset{(c)}{\le} \varepsilon_i - \sigma^c_{min} - \log \frac{e^{-\varepsilon_i + \sigma^c_{min}} \cdot \sum_{x \in \mathcal{D}^n} p_{Y|X}(y \mid x') p_X(x)}{p_{Y|X}(y \mid x')} \tag{2.31}$$

$$= \varepsilon_i - \sigma^c_{min} - \log(e^{-\varepsilon_i + \sigma^c_{min}} \cdot \sum_{x \in \mathcal{D}^n} p_X(x)) \tag{2.32}$$

$$= 2(\varepsilon_i - \sigma^c_{min}), \tag{2.33}$$

where (a) follows the fact that

$$\log \frac{p_{X|Y}(x \mid y)}{p_{X|Y}(x' \mid y)} \tag{2.34}$$

$$= \log \left(\frac{p_{X,Y}(x, y)}{p_X(x)p_Y(y)} \cdot \frac{p_X(x)}{p_X(x')} \cdot \frac{p_X(x')p_Y(y)}{p_{X,Y}(x', y)} \right) \tag{2.35}$$

$$= \log \frac{p_{X,Y}(x, y)}{p_X(x)p_Y(y)} + \log \frac{p_X(x)}{p_X(x')} + \log \frac{p_Y(y)}{p_{Y|X}(y \mid x')} \tag{2.36}$$

$$= \log \frac{p_{X,Y}(x, y)}{p_X(x)p_Y(y)} + \log \frac{p_X(x)}{p_X(x')} + \log \frac{\sum_{x \in \mathcal{D}^n} p_{Y|X}(y \mid x)p_X(x)}{p_{Y|X}(y \mid x')}, \tag{2.37}$$

(b) follows Eq. (2.7), the definition of minimum prior probability difference between two neighboring datasets, i.e.,

$$-\log \frac{p_X(x)}{p_X(x')} \leq -\sigma^c_{min}, \ \forall x, x' \in \mathcal{D}^n : x \sim x', \tag{2.38}$$

and (c) follows Theorem 2.1, i.e., ε_d-central identifiability implies $(\varepsilon_d - \sigma^c_{min})$-CDP,

$$e^{-\varepsilon_i + \sigma^c_{min}} \cdot p_{Y|X}(y \mid x') \leq p_{Y|X}(y \mid x), \ \forall x, x' \in \mathcal{D}^n : x \sim x'. \tag{2.39}$$

As a result, we have

$$\sum_{x,y \in \mathcal{D}^n} p_{X,Y}(x, y) \log \frac{p_{X,Y}(x, y)}{p_X(x)p_Y(y)} \tag{2.40}$$

$$\leq 2(\varepsilon_i - \sigma^c_{min}) \cdot \sum_{x,y \in \mathcal{D}^n} p_{X,Y}(x, y) \tag{2.41}$$

$$= 2(\varepsilon_i - \sigma^c_{min}). \tag{2.42}$$

Therefore, the mechanism $\mathcal{A} := p_{Y|X}(y \mid x)$ satisfies $2(\varepsilon_i - \sigma^c_{min})$-mutual-information privacy. Similarly, we can prove that the mechanism $\mathcal{A} := p_{Y|X}(y \mid x)$ satisfying ε_i-local identifiability also satisfies ε_m-lcoal mutual-information privacy with $\varepsilon_m = 2(\varepsilon_i - \sigma^l_{min})$. In summary, ε_i-identifiability implies $2(\varepsilon_i - \sigma_{min})$-mutual-information privacy. □

Theorem 2.4 (Mutual-Information Privacy does not Guarantee Identifiability) *The mechanism $\mathcal{A}$ that satisfies ε_m-mutual-information privacy does not imply ε_i-identifiability, i.e., the value of ε_i is not related to ε_m.*

Proof In the central scenario, we demonstrate through the following special case that mutual-information privacy cannot guarantee identifiability. Given $\mathcal{D}^2 = \{(1, 1), (1, 2), (2, 2)\}$ and the joint probability distribution $p_{X,Y}$ is

$$p_{X,Y}(x, y) = \begin{cases} \beta, & x = y = (1, 1), \\ 0, & x = (1, 1),\ \forall y \in \mathcal{D}^2, \\ 0, & \forall x \in \mathcal{D}^2,\ y = (1, 1), \\ \dfrac{1-\beta}{4}, & \text{others}, \end{cases} \tag{2.43}$$

where $0 < \beta < 1$. In this case, the mutual information between X and Y can be formulated as

$$I(X; Y) = \sum_{x,y \in \mathcal{D}^2} p_{X,Y}(x, y) \log \frac{p_{X,Y}(x, y)}{p_X(x)p_Y(y)} \tag{2.44}$$

$$= \beta \cdot \log \frac{1}{\beta} + (1 - \beta) \cdot \log \frac{1}{1 - \beta}. \tag{2.45}$$

Thus, the mechanism $\mathcal{A} := p_{Y|X}(y \mid x)$ satisfies ε_m-mutual-information privacy with bounded constant $\varepsilon_m = \beta \cdot \log(1/\beta) + (1 - \beta) \cdot \log(1/(1 - \beta))$. Then, we assume that the mechanism $\mathcal{A} := p_{Y|X}(y \mid x)$ satisfies ε_i-identifiability, where the privacy budget ε_i can be formulated as

$$\varepsilon_i = \max_{x,x',y \in \mathcal{D}^2 : x \sim x'} \log \frac{p_{X|Y}(x \mid y)}{p_{X|Y}(x' \mid y)} \tag{2.46}$$

$$= \max_{x,x',y \in \mathcal{D}^2 : x \sim x'} \log \frac{p_{X,Y}(x, y)}{p_{X,Y}(x', y)} \tag{2.47}$$

$$= +\infty. \tag{2.48}$$

However, the unbounded privacy budget ε_i means that the mechanism $\mathcal{A} := p_{Y|X}(y \mid x)$ does not guarantee ε_i-identifiability.

In the local scenario, similarly, we demonstrate that mutual-information privacy cannot guarantee identifiability through the following special case. Given $\mathcal{D} = \{1, 2\}$ and the joint probability distribution $p_{D,E}$ is

$$p_{D,E}(d, e) = \begin{cases} \beta, & d = e = 1, \\ 1 - \beta, & d = e = 2, \\ 0, & \text{others}, \end{cases} \tag{2.49}$$

where $0 < \beta < 1$. According to Definition 2.7, we conclude that the mechanism $\mathcal{A} := p_{E|D}(e \mid d)$ satisfies ε_m-mutual-information privacy, where privacy budget ε_m can be formulated as

$$\varepsilon_m = I(D; E) \tag{2.50}$$

$$= \sum_{d,e \in \mathcal{D}} p_{D,E}(d, e) \log \frac{p_{D,E}(d, e)}{p_D(d) p_E(e)} \tag{2.51}$$

$$= \beta \cdot \log \frac{1}{\beta} + (1 - \beta) \cdot \log \frac{1}{1 - \beta}. \tag{2.52}$$

However, the mechanism $\mathcal{A} := p_{E|D}(e \mid d)$ cannot satisfies ε_i-identifiability since the privacy budget $\varepsilon_i = +\infty$, i.e.,

$$\varepsilon_i = \max_{d_i, d_j \in \mathcal{D}} \log \frac{p_{D|E}(d_j \mid e)}{p_{D|E}(d_k \mid e)} \tag{2.53}$$

$$= +\infty. \tag{2.54}$$

In summary, mutual-information privacy does not guarantee identifiability. □

2.2.3 *Differential Privacy vs. Mutual-Information Privacy*

DP defines the indistinguishability between any neighboring datasets in the finite domain $\mathcal{D}^n$ (or any data in the finite domain $\mathcal{D}$), while mutual-information privacy measures the average amount of information contained in the perturbed dataset y (or the perturbed data e) that reveals the real dataset x (or the real data d). According to definitions of DP, mutual-information privacy, and minimum prior probability difference, we prove that the mechanism $\mathcal{A}$ satisfying ε_d-DP also satisfies ε_m-mutual-information privacy with $\varepsilon_m = 2\varepsilon_d$, and the mechanism $\mathcal{A}$ satisfying ε_m-mutual-information privacy does not guarantee ε_d-DP. The detailed theorems and corresponding proofs are given below.

Theorem 2.5 **(ε_d-Differential Privacy Implies $2\varepsilon_d$-Mutual-Information Privacy)** *The mechanism $\mathcal{A}$ also satisfies ε_i-mutual-information privacy with $\varepsilon_i = 2\varepsilon_d$ when it satisfies ε_d-DP.*

Proof In the central scenario, since the mechanism $\mathcal{A} := p_{Y|X}(y \mid x)$ satisfies ε_d-CDP, as defined in Definition 2.4, we have

$$\frac{p_{Y|X}(y \mid x)}{p_{Y|X}(y \mid x')} \leq e^{\varepsilon_d}, \ \forall x, x' \in \mathcal{D}^n : x \sim x'. \tag{2.55}$$

Taking the logarithm of both sides of Eq. (2.55) yields

$$\log \frac{p_{Y|X}(y \mid x)}{p_{Y|X}(y \mid x')} \leq \varepsilon_d, \ \forall x, x' \in \mathcal{D}^n : x \sim x'. \tag{2.56}$$

Besides, Eq. (2.56) can be reformulated as

$$\log \frac{p_{X,Y}(x,y)}{p_X(x)p_Y(y)} \leq \varepsilon_d - \log \frac{\sum_{x\in\mathcal{D}^n} p_{Y|X}(y \mid x)p_X(x)}{p_{Y|X}(y \mid x')}, \ \forall x, x' \in \mathcal{D}^n : x \sim x'. \tag{2.57}$$

This is due to the fact that

$$\log \frac{p_{Y|X}(y \mid x)}{p_{Y|X}(y \mid x')} \tag{2.58}$$

$$= \log \left(\frac{p_{X,Y}(x,y)}{p_X(x)p_Y(y)} \cdot \frac{p_X(x')p_Y(y)}{p_{X,Y}(x',y)} \right) \tag{2.59}$$

$$= \log \left(\frac{p_{X,Y}(x,y)}{p_X(x)p_Y(y)} \cdot \frac{p_X(x')p_Y(y)}{p_{Y|X}(y \mid x')p_X(x')} \right) \tag{2.60}$$

$$= \log \left(\frac{p_{X,Y}(x,y)}{p_X(x)p_Y(y)} \cdot \frac{\sum_{x\in\mathcal{D}^n} p_{Y|X}(y \mid x)p_X(x)}{p_{Y|X}(y \mid x')} \right) \tag{2.61}$$

$$= \log \frac{p_{X,Y}(x,y)}{p_X(x)p_Y(y)} + \log \frac{\sum_{x\in\mathcal{D}^n} p_{Y|X}(y \mid x)p_X(x)}{p_{Y|X}(y \mid x')}. \tag{2.62}$$

According to the equivalent form of CDP, i.e.,

$$e^{-\varepsilon_d} \cdot p_{Y|X}(y \mid x') \leq p_{Y|X}(y \mid x), \ \forall x, x' \in \mathcal{D}^n : x \sim x', \tag{2.63}$$

we have

$$\log \frac{\sum_{x\in\mathcal{D}^n} p_{Y|X}(y \mid x)p_X(x)}{p_{Y|X}(y \mid x')} \tag{2.64}$$

$$\geq \log \frac{e^{-\varepsilon_d} \cdot \sum_{x\in\mathcal{D}^n} p_{Y|X}(y \mid x')p_X(x)}{p_{Y|X}(y \mid x')} \tag{2.65}$$

$$= \log(e^{-\varepsilon_d} \cdot \sum_{x\in\mathcal{D}^n} p_X(x)) \tag{2.66}$$

$$= -\varepsilon_d. \tag{2.67}$$

Powered by Eq. (2.67), Eq. (2.57) can be reformulated as

$$\log \frac{p_{X,Y}(x,y)}{p_X(x)p_Y(y)} \leq 2\varepsilon_d. \tag{2.68}$$

As a result, we have

$$\sum_{x,y\in\mathcal{D}^n} p_{X,Y}(x,y)\log\frac{p_{X,Y}(x,y)}{p_X(x)p_Y(y)} \tag{2.69}$$

$$\leq 2\varepsilon_d\cdot\sum_{x,y\in\mathcal{D}^n} p_{X,Y}(x,y) \tag{2.70}$$

$$=2\varepsilon_d, \tag{2.71}$$

that is, the mechanism $\mathcal{A} := p_{Y|X}(y \mid x)$ satisfies $2\varepsilon_d$- central mutual-information privacy. Similarly, we can prove that the mechanism $\mathcal{A} := p_{E|D}(e \mid d)$ satisfying ε_d-LDP also satisfies ε_m-local mutual-information privacy with $\varepsilon_m = 2\varepsilon_d$. In summary, ε_d-DP implies $2\varepsilon_d$-mutual-information privacy. □

Theorem 2.6 (Mutual-Information Privacy does not Guarantee Differential Privacy) *The mechanism $\mathcal{A}$ that satisfies ε_m-mutual-information privacy does not imply ε_d-DP, i.e., the value of ε_d is not related to ε_m.*

Proof In the central scenario, we demonstrate through the following special case that mutual-information privacy cannot guarantee DP. Given $\mathcal{D}^2 = \{(1,1),(1,2),(2,2)\}$ and the joint probability distribution $p_{X,Y}$ is

$$y_i = \begin{cases} \beta, & x = y = (1,1), \\ 0, & x = (1,1),\ \forall y \in \mathcal{D}^2, \\ 0, & \forall x \in \mathcal{D}^2,\ y = (1,1), \\ \dfrac{1-\beta}{4}, & \text{others}, \end{cases} \tag{2.72}$$

where $0 < \beta < 1$. Since the mechanism $\mathcal{A} := p_{Y|X}(y \mid x)$ satisfies ε_m-mutual-information privacy, the privacy budget ε_m can be formulated as

$$\varepsilon_m = I(X;Y) \tag{2.73}$$

$$= \sum_{x,y\in\mathcal{D}^2} p_{X,Y}(x,y)\cdot\log\frac{p_{X,Y}(x,y)}{p_X(x)p_Y(y)} \tag{2.74}$$

$$= \beta\cdot\log\frac{1}{\beta} + (1-\beta)\cdot\log\frac{1}{1-\beta}. \tag{2.75}$$

Obviously, the privacy budget ε_m is a bounded constant. In the above specific case, we assume that the mechanism $\mathcal{A} := p_{Y|X}(y \mid x)$ satisfying ε_m-central mutual-information privacy also satisfies ε_d-CDP. Then, according to the definition of CDP (detailed in Definition 2.4), the privacy budget ε_d can be formulated as

$$\varepsilon_d = \max_{x,x',y\in\mathcal{D}^2:x\sim x'} \log \frac{p_{Y|X}(y \mid x)}{p_{Y|X}(y \mid x')} \tag{2.76}$$

$$= \max_{x,x',y\in\mathcal{D}^2:x\sim x'} \log \left(\frac{p_X(x')}{p_X(x)} \cdot \frac{p_{X,Y}(x, y)}{p_{X,Y}(x', y)} \right) \tag{2.77}$$

$$= +\infty. \tag{2.78}$$

However, the privacy budget ε_d is unbounded, indicating that the assumption regarding the mechanism $\mathcal{A} := p_{Y|X}(y \mid x)$ satisfies ε_d-DP does not hold. As a result, the mechanism $\mathcal{A} := p_{Y|X}(y \mid x)$ that satisfies ε_m-mutual-information privacy does not guarantee ε_d-DP. Similarly, we demonstrate through the following special case that mutual-information privacy cannot guarantee DP in the local scenario. Given $\mathcal{D} = \{1, 2\}$ and the joint probability distribution $p_{D,E}$ is

$$p_{D,E}(d, e) = \begin{cases} \beta, & d = e = 1, \\ 1 - \beta, & d = e = 2, \\ 0, & \text{others}, \end{cases} \tag{2.79}$$

where $0 < \beta < 1$. According to the Definition 2.7, we can obtain that the mechanism $\mathcal{A} := p_{E|D}$ satisfies ε_m-local mutual-information privacy, where privacy budget ε_m can be formulated as

$$\varepsilon_m = I(D; E) \tag{2.80}$$

$$= \beta \cdot \log \frac{1}{\beta} + (1 - \beta) \cdot \log \frac{1}{1 - \beta}. \tag{2.81}$$

However, the mechanism $\mathcal{A} := p_{E|D}(e \mid d)$ cannot satisfies ε_d-LDP, since

$$\varepsilon_d = \max_{d_i,d_j,e\in\mathcal{D}} \log \frac{p_{D|E}(e \mid d_j)}{p_{D|E}(e \mid d_k)} \tag{2.82}$$

$$= \max_{d_i,d_j,e\in\mathcal{D}} \log \left(\frac{p_D(d_j)}{p_D(d_i)} \cdot \frac{p_{D,E}(d_i, e)}{p_{D,E}(d_j, e)} \right) \tag{2.83}$$

$$= +\infty. \tag{2.84}$$

In summary, mutual-information privacy does not guarantee DP. □

2.3 Summary

In this chapter, we have given several classical privacy notions, including identifiability, differential privacy, and mutual-information privacy, for achieving privacy preservation in crowdsensing systems. Besides, we have proposed a completeness framework, linking those privacy notions. This framework constitutes the cornerstone for perturbation-based privacy in crowdsensing systems in the subsequent chapters.

References

1. Zheng, Z., Li, Z., Li, J., Jiang, H., Li, T., Guo, B.: Utility-aware and privacy-preserving trajectory synthesis model that resists social relationship privacy attacks. ACM Trans. Intel. Syst. Technol. **13**(3), 1–28 (2022)
2. Zheng, Z., Li, Z., Jiang, H., Zhang, L.Y., Tu, D.: Semantic-aware privacy-preserving online location trajectory data sharing. IEEE Trans. Inf. Foren. Secur. **17**, 2256–2271 (2022)
3. Zheng, Z., Li, Z., Huang, C., Long, S., Shen, X.: Defending data poisoning attacks in dp-based crowdsensing: a game-theoretic approach. IEEE Trans. Mobile Comput. **24**(3), 1859–1876 (2024)
4. Zheng, Z., Li, Z., Huang, C., Long, S., Li, M., Shen, X.: Data poisoning attacks and defenses to LDP-based privacy-preserving crowdsensing. IEEE Trans. Depend. Secur. Comput. **21**(5), 4861–4878 (2024)
5. Li, Z., Zheng, Z., Guo, S., Guo, B., Xiao, F., Ren, K.: Disguised as privacy: data poisoning attacks against differentially private crowdsensing systems. IEEE Trans. Mobile Comput. **22**(9), 5155–5169 (2023)
6. Wang, T., Zheng, Z., Rehmani, M.H., Yao, S., Huo, Z.: Privacy preservation in big data from the communication perspective-a survey. IEEE Commun. Surv. Tuts. **21**(1), 753–778 (2019)
7. Wang, T., Chen, J.Q., Zhang, Z., Su, D., Cheng, Y., Li, Z., Li, N., Jha, S.: Continuous release of data streams under both centralized and local differential privacy. In: Proceedings of ACM SIGSAC Conference on Computer and Communications Security, pp. 1237–1253 (2021)
8. Fallah, A., Makhdoumi, A., Malekian, A., Ozdaglar, A.: Bridging central and local differential privacy in data acquisition mechanisms. In: Proceedings of Advances in Neural Information Processing Systems, vol. 35, pp. 21628–21639 (2022)
9. Wu, L., Qin, C., Xu, Z., Guan, Y., Lu, R.: TCPP: achieving privacy-preserving trajectory correlation with differential privacy. IEEE Trans. Inf. Foren. Secur. **18**, 4006–4020 (2023)
10. Huang, Y., Xue, K., Zhu, B., Zhao, J., Li, R.: Collecting partial ordered data with local differential privacy. IEEE Trans. Inf. Foren. Secur. **19**, 7646–7658 (2024)
11. Zheng, Z., Li, Z., Long, S., Guo, S., Chen, C., Xu, K.: Pricing utility vs. location privacy: a differentially private data sharing framework for ride-on-demand services. IEEE Trans. Depend. Secur. Comput. 1–17 (2025). https://doi.org/10.1109/TDSC.2025.3532599
12. Yang, M., Guo, T., Zhu, T., Tjuawinata, I., Zhao, J., Lam, K.Y.: Local differential privacy and its applications: a comprehensive survey. Comput. Stand. Interfaces **89**, 103827 (2024)
13. Dwork, C., McSherry, F., Nissim, K., Smith, A.: Calibrating noise to sensitivity in private data analysis. In: Proceedings of Theory of Cryptography: Third Theory of Cryptography Conference, pp. 265–284 (2006)
14. Dwork, C.: Differential privacy. In: Proceedings of International Colloquium on Automata, Languages, and Programming, pp. 1–12 (2006)

15. Duchi, J.C., Jordan, M.I., Wainwright, M.J.: Local privacy and statistical minimax rates. In: Proceedings of IEEE Annual Symposium on Foundations of Computer Science, pp. 429–438 (2013)
16. Zhao, Y., Du, J.T., Chen, J.: Scenario-based adaptations of differential privacy: a technical survey. ACM Comput. Surv. **56**(8), 1–39 (2024)
17. Lee, J., Clifton, C.: Differential identifiability. In: Proceedings of ACM SIGKDD International Conference on Knowledge Discovery and Data Mining, pp. 1041–1049 (2012)
18. Wang, W., Ying, L., Zhang, J.: On the relation between identifiability, differential privacy, and mutual-information privacy. IEEE Trans. Inf. Theory **62**(9), 5018–5029 (2016)
19. McGraw, D.: Data identifiability and privacy. Am. J. Bioethics **10**(9), 30–31 (2010)
20. Sankar, L., Rajagopalan, S.R., Poor, H.V.: Utility-privacy tradeoffs in databases: an information-theoretic approach. IEEE Trans. Inf. Foren. Secur. **8**(6), 838–852 (2013)
21. Cuff, P., Yu, L.: Differential privacy as a mutual information constraint. In: Proceedings of ACM SIGSAC Conference on Computer and Communications Security, pp. 43–54 (2016)
22. Liao, J., Sankar, L., Tan, V.Y.F., du Pin Calmon, F.: Hypothesis testing under mutual information privacy constraints in the high privacy regime. IEEE Trans. Inf. Foren. Secur. **13**(4), 1058–1071 (2018)
23. Wang, B., Guo, J., Li, A., Chen, Y., Li, H.: Privacy-preserving representation learning on graphs: a mutual information perspective. In: Proceedings of ACM SIGKDD Conference on Knowledge Discovery & Data Mining, pp. 1667–1676 (2021)
24. Oya, S., Troncoso, C., Pérez-González, F.: Back to the drawing board: revisiting the design of optimal location privacy-preserving mechanisms. In: Proceedings of ACM SIGSAC Conference on Computer and Communications Security, pp. 1959–1972 (2017)
25. Latham, P.E., Roudi, Y.: Mutual information. Scholarpedia **4**(1), 1658 (2009)

Chapter 3
Semantic-Aware Trajectory Privacy Preservation in Crowdsensing

3.1 Introduction

With the growth of crowdsensing, users can access numerous services by sharing their location information [1]. For instance, drivers receive navigation services through Google Maps, and users can find popular nearby restaurants via Meituan. Moreover, crowdsensing servers can collect and publish users' trajectory datasets over time to support applications such as intelligent transportation [2], urban computing [3], and mobile service provisioning [4, 5]. However, releasing trajectory datasets raises significant privacy concerns due to the sensitive nature of location data. As reported in [6], merely four spatiotemporal points can uniquely identify about 95% of individuals in a detailed trajectory dataset. Therefore, protecting the privacy of shared location trajectory data is essential.

Existing location privacy protection techniques are primarily designed for trajectory data publishing and can be broadly divided into offline [7–15] and online [16–26] scenarios. The key difference is that online methods do not have access to whole trajectories, which makes offline techniques like trajectory synthesis [10, 11] and merging [13] inapplicable. Recently, information-theoretic privacy (e.g., mutual-information privacy) have emerged for online scenarios [17, 21–23, 27], allowing users to sanitize their current location without requiring complete trajectory data. As a result, users can access crowdsensing services with minimal impact on service quality (i.e., utility) while their sensitive information (i.e., privacy) remains protected.

In centralized online trajectory privacy solutions, a third party is typically employed to execute privacy protection mechanisms. However, these solutions are prone to vulnerabilities since such third parties can be targets of malicious attacks or may compromise users' privacy in pursuit of commercial benefits. Notably, breaches like those involving Facebook and Cambridge Analytica have demonstrated the severe consequences of these weaknesses, resulting in substantial financial losses [28]. Therefore, it is crucial to develop a user-centric privacy-

Z. Zheng et al., *Perturbation Based Privacy in Crowdsensing*, Wireless Networks,
https://doi.org/10.1007/978-3-031-95052-0_3

preserving approach that operates locally without relying on any external trusted party.

However, a key issue is that many current mechanisms for privacy-preserving online trajectory sharing perturb each location individually without accounting for the spatiotemporal correlations among locations within a trajectory [17, 21, 22, 27]. Here, data privacy means that a semi-honest (honest but curious) server should be unable to deduce the real location from the perturbed data. However, neglecting the correlations within a trajectory leaves these methods vulnerable, since the server can exploit temporal and spatial constraints (e.g., time reachability) to rule out implausible locations or even recover the real location [29]. For example, if a user is on a highway, they cannot suddenly move perpendicularly off the road, a fact that an adversary could detect by analyzing previous locations.

Moreover, merely protecting data privacy does not suffice to preserve semantic privacy regarding the purpose of visiting. We define the Point of Interest (PoI) distribution as the probability distribution representing the ratio of each PoI category to the overall count. When locations are perturbed, the resulting PoI distribution may still allow the semi-honest server to infer the underlying motivation for visiting, leading to semantic privacy leakage [13, 30]. For example, even if Bob masks his exact university location, the surrounding PoIs remain predominantly university-related, making it easy for adversaries to deduce that he is likely a student. The similarity between the PoI distributions of real and perturbed locations thus highlights the vulnerability of semantic privacy.

Although the semantic features embedded in trajectories can inadvertently expose users' visit purposes, they are also essential for delivering personalized services [31, 32]. For example, while Alice and Ruth both traveled from Beijing to Shanghai, Alice's route mainly featured toll roads, whereas Ruth's path included free roads with scenic lake views. Despite the spatial proximity of their routes, these differences enable the server to provide tailored travel recommendations in the future. Consequently, there is an urgent need for a user-centric online mechanism for sharing location trajectory data, protecting both data and semantic privacy without sacrificing the valuable semantic features necessary for personalized services.

Motivated by the above discussions, we focus on sharing online location trajectory data with a semi-honest server. Our goal is to protect users' data and semantic privacy while retaining the data's semantic utility. This setting presents three key challenges. *The first is accurately quantifying privacy leakage.* The spatiotemporal correlations among locations within a trajectory and the causal correlations between real and perturbed trajectories contribute to data privacy leakage, and discrepancies in PoI distributions between real and perturbed locations lead to semantic privacy leakage. Capturing these correlations is challenging because, in online scenarios, we lack access to the complete trajectory. *The second challenge is protecting both data and semantic privacy.* Protecting data privacy does not automatically prevent inferences about a user's visit purpose, making it difficult to design a mechanism that protects both aspects simultaneously. *The third challenge is balancing privacy and semantic-aware utility.* Preserving semantic features inherently involves some privacy compromise, yet excessive perturbation, while enhancing privacy, can severely diminish data utility. Therefore, finding the optimal trade-off between these conflicting goals is a critical and nontrivial task.

To overcome the challenges, we introduce SEmantic-aware Information-Theoretic Privacy (SEITP), a novel mechanism for privacy-preserving online location trajectory data sharing. SEITP protects both data and semantic privacy while ensuring semantic-aware data utility. To quantify privacy and utility, we define three key metrics based on information theory: a trajectory-level data privacy metric, a semantic privacy metric, and a semantic-aware utility metric. Then, we formulate the problem as a multi-objective optimization problem, achieving an optimal balance between privacy and utility. By solving this optimization problem, we derive a two-stage local privacy mapping SEITP to protect both data and semantic privacy. The key contributions of this chapter are as follows:

- We propose trajectory-level data privacy and semantic privacy metrics based on information theory to quantify data and semantic privacy leakage, respectively. Building on these metrics, we design a novel two-stage privacy-preserving method and provide theoretical proof of its effectiveness in protecting both data and semantic privacy.
- Utilizing the newly proposed metrics, we formulate the optimal online location trajectory data sharing mechanism as a multi-objective optimization problem that seeks to minimize privacy leakage while maintaining a predefined utility level. To achieve the optimal two-stage local privacy mapping (i.e., SEITP), we decompose the multi-objective optimization problem into a data privacy sub-problem and a semantic privacy sub-problem. These are then efficiently solved using the alternating direction multiplier method and the artificial bee colony algorithm, respectively.
- We evaluate SEITP by comparing it with state-of-the-art methods using a real-world private vehicle trajectory dataset. Experimental results demonstrate that SEITP effectively protects both data and semantic privacy while preserving superior data utility.

The remainder of this chapter is structured as follows. Section 3.2 defines the problem and introduces fundamental concepts. Sections 3.3 and 3.4 elaborate on the SEITP mechanism. The theoretical analysis is provided in Sect. 3.5. Section 3.6 discusses performance evaluation and experimental results. Finally, Sect. 3.7 summarizes this chapter.

3.2 Problem Statement and Basic Concepts

3.2.1 Problem Statement

We state the problem addressed in this chapter from the following three aspects: the system model, the adversary model, and the design goals.

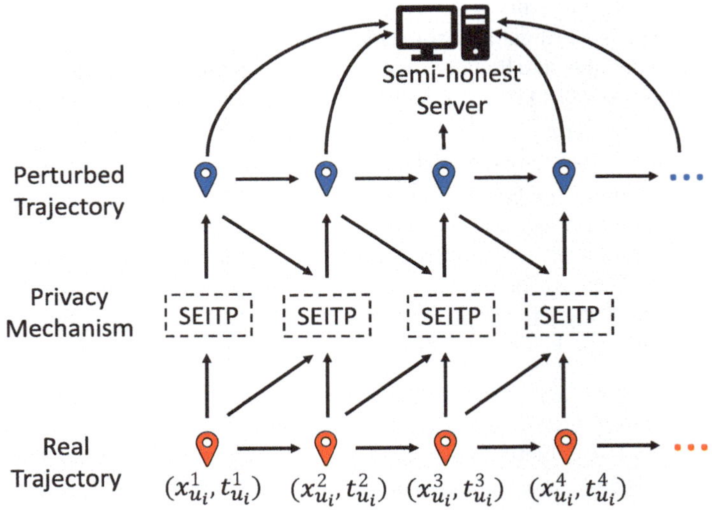

Fig. 3.1 Diagram illustrating privacy-preserving online sharing of location trajectories, where spatiotemporal correlations among locations within a trajectory are captured using a first-order Markov chain

System Model We consider the system model shown in Fig. 3.1. Rather than revealing the real location $x_{u_i}^j$, a privacy-aware user u_i uploads a perturbed version $z_{u_i}^j$ to a crowdsensing server at each timestamp $t_{u_i}^j$ for location services. In real time, u_i generates $z_{u_i}^j$ using the privacy-preserving mechanism SEITP before sharing it with the server. Over time, the server aggregates these reports to construct the user's trajectory, which is later published for academic or commercial use. For simplicity, since this chapter focuses on a single user, we omit the subscript u_i and denote $x^j := x_{u_i}^j$, $y^j := y_{u_i}^j$, $z^j := z_{u_i}^j$, and $t^j := t_{u_i}^j$.

Building on prior research [17, 21–23, 27], we consider the server as semi-honest. This means the server reliably carries out geo-data analysis to deliver high-quality services while potentially probing users' sensitive information and, driven by profit, possibly sharing trajectory data with advertisers or illegal organizations.

In real-world crowdsensing systems, users frequently broadcast their location data. For instance, a user must provide their current location to Google Maps for navigation, and Didi Taxi continuously updates location information to support ride-hailing services. Importantly, this chapter focuses on locally generating perturbed locations, eliminating the need for any third party to execute SEITP.

Adversary Model As illustrated in Fig. 3.1, after a certain period the semi-honest server collects the perturbed trajectory $Z_i(T) = \{(z^1, t^1), (z^2, t^2), \cdots, (z^T, t^T)\}$. Based on this data, the server may infer sensitive information about the user. We assume that the adversary has prior knowledge of the victim's mobility behavior, specifically the transition probability matrix defined over grid cells. In addition, the attacker possesses auxiliary information such as the overall PoI distribution of the city. Essentially, by combining the transition probability matrix with the perturbed trajectory, the adversary can launch various inference attacks (e.g., location-dependent attacks [19]), which can lead to data privacy leaks. Moreover, the adversary can use both the PoI distribution and the perturbed trajectory to carry out semantic attacks [13, 30] in order to uncover the motivation behind visiting real locations, thereby causing semantic privacy leakage.

Design Goals Our objective is to ensure that the crowdsensing server both protects users' data and semantic privacy and delivers high-quality location-based services. To achieve this, we must assess the extent of privacy leakage that arises from sharing perturbed location trajectories, and quantify the degradation in service quality resulting from noise perturbation.

3.2.2 *Basic Concepts*

Before introducing the semantic-aware location trajectory privacy mapping mechanism $\mathcal{A}$, we outline several fundamental concepts: mutual information [33], directed information [34], transition probability matrix, and PoI distribution.

Mutual Information Mutual information [33] quantifies the degree of dependence between two random variables, and its formal definition is presented below.

Definition 3.1 (Mutual Information) Let (X, Y) be a pair of random variables, the mutual information between X and Y is defined as

$$I(X; Y) = H(X) - H(X \mid Y), \tag{3.1}$$

where $H(X \mid Y)$ is the conditional entropy.

Directed Information Directed information [34] quantifies how much information is transferred from one random process to another. Below, we provide its detailed definition.

Definition 3.2 (Directed Information) Let $\mathbf{X}$ denote the vector $(X_1, X_2, \cdots, X_n)$ and $\mathbf{Y}$ denote $(Y_1, Y_2, \cdots, Y_n)$, the directed information from $\mathbf{X}$ to $\mathbf{Y}$ is defined as

$$I(\mathbf{X} \to \mathbf{Y}) = \sum_{i=1}^{n} I(X^{i:1}; Y_i \mid Y^{i-1:1}) \tag{3.2}$$

$$= \sum_{i=1}^{n} I(X^{i:1}; Y_i) - I(X^{i:1}; Y_i; Y^{i-1:1}), \tag{3.3}$$

where $X^{i:1} = (X_1, X_2, \cdots, X_i)$.

Mobility Model We model the user mobility as a Markov chain of order r on the set of regions (locations). Consequently, a user u's mobility model is characterized by the transition probability matrix of this chain, whose detailed definition follows.

Definition 3.3 (Transition Probability Matrix) Let $p(u)$ be a transition probability matrix of user u. The transition probability matrix $p(u)$ is formulated as

$$p(u) = \left(p(x^k = l_1 \mid x^{k-1:k-r}), \cdots, p(x^k = l_D \mid x^{k-1:k-r})\right), \tag{3.4}$$

where l_m is real location, D denotes the number of spatial grid, and $\sum_{m=1}^{D} p(x^k = l_m \mid x^{k-1:k-r}) = 1$. Specifically, if $k <= r$, we have $p(x^k = l_m \mid x^{k-1:k-r}) = p(x^k = l_m \mid x^{k-1:1})$.

PoI Distribution PoI is a feature occupying a particular location on the map, which reveals most of the semantic information of a location. Since each location contains several PoIs, we can give the PoI distribution of each location. The detailed definition of the PoI distribution is given below.

Definition 3.4 (PoI Distribution) Let $P(x^j)$ denote the PoI distribution of location x^j. The PoI distribution $P(x^j)$ is formulated as

$$P(x^j) = (p_1, p_2, \cdots, p_n), \tag{3.5}$$

where the element p_i ($i \in [1, n]$ and $\sum_{i=1}^{n} p_i = 1$) represents the ratio of the number of category i PoI (e.g., number of food stores) to the total number of PoIs of all categories in a circular area centered at the location x^j.

3.3 Privacy and Utility Metrics

Although perturbation-based privacy-preserving mechanisms protect users' privacy, the added noise unavoidably compromises data utility. Consequently, we propose two distinct privacy metrics, a trajectory-level data privacy metric and a semantic privacy metric, to assess the extent to which sharing perturbed data exposes real locations and visit purposes. We also propose a semantic-aware utility metric to evaluate the loss of data utility resulting from noise perturbation.

3.3.1 Data Privacy Metric

Existing research offers numerous privacy metrics for quantifying data leakage in online single-location sharing scenarios [17, 21, 22, 27], such as mutual information

$I(x^j; z^j)$. However, these metrics prove inadequate for trajectory data sharing as they neglect spatiotemporal correlations among locations within a trajectory. To address this limitation, trajectory-level privacy metrics like $I(X_i(T); Z_i(T))$ have emerged in recent years [23], aiming to capture holistic privacy risks in continuous location sharing. Nevertheless, current approaches fundamentally misinterpret the interaction between real trajectories $X_i(T)$ and perturbed outputs $Z_i(T)$ by modeling them as symmetrical mutual influences. This assumption contradicts the inherent causal structure of trajectory privacy protection mechanisms. From a user perspective, the protection process follows a spatiotemporal causal chain, i.e., given historical real locations $\{x^1, x^2, \cdots, x^t\}$ and previously released perturbed locations $\{z^1, z^2, \cdots, z^{t-1}\}$, the user strategically generates the current z^t (effect) to minimize leakage of causal data $\{x^1, x^2, \cdots, x^t\}$. Conventional undirected correlation models fail to distinguish this cause-effect correlation, leading to inaccurate quantification of privacy leaks. This critical gap motivates our proposal to adopt directed information [34] for developing causality-aware trajectory privacy metrics that properly model spatiotemporal correlations and unidirectional information flow in location sharing processes.

Based on Definition 3.2, we define the directed information from the real trajectory $X_i(T)$ to perturbed trajectory $Z_i(T)$ as follows,

$$I\left(X_i(T) \to Z_i(T)\right) := \sum_{j=1}^{T} I\left(x^{j:1}; z^j \mid z^{j-1:1}\right), \tag{3.6}$$

where $I(x^{j:1}; z^j \mid z^{j-1:1})$ represents conditional mutual information and $x^{j:j-k} = \{x^j, \cdots, x^{j-k}\}$ for $0 \le k < j$. Specifically, we have $I(x^{j:1}; z^j \mid z^{j-1:1}) = I(x^1; z^1)$ when $j = 1$. Fundamentally, directed information $I(X_i(T) \to Z_i(T))$ quantifies the causal influence of the real trajectory $X_i(T)$ on the perturbed trajectory $Z_i(T)$ [34, 35], capturing how historical locations of $X_i(T)$ sequentially shape the generation of $Z_i(T)$ over time (as shown in Fig. 3.2). In summary, we define the trajectory-level data privacy metric in detail as follows.

Definition 3.5 (Trajectory-Level Data Privacy Metric) Let $X_i(T)$ and $Z_i(T)$ be the real and perturbed trajectory of user u_i. Assume the user's real trajectory follow a Markov chain of order r, i.e., real location x^j depends only on the real location sequence $\{x^{j-1}, x^{j-2}, \cdots, x^{j-r}\}$. We also assume that the user's perturbed trajectory follows a multivariate Markov chain [36] of order r, i.e., perturbed location z^j depends only on the location sequence $\{x^{j-1}, \cdots, x^{j-r}, z^{j-1}, \cdots, z^{j-r}\}$. The trajectory-level data privacy metric is defined as

$$\mathcal{D}(X_i(T); Z_i(T)) = I\left(X_i(T) \to Z_i(T)\right) \tag{3.7}$$

$$\overset{(a)}{=} \sum_{j=1}^{T} I\left(x^j; z^{T:j} \mid x^{j-1:1}, z^{j-1:1}\right) \tag{3.8}$$

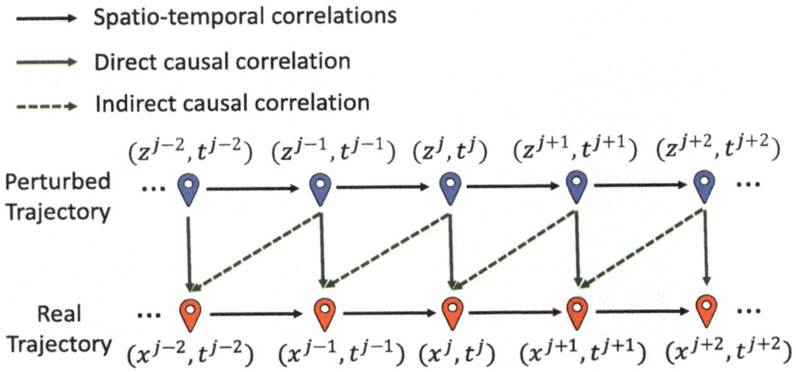

Fig. 3.2 The correlations of real locations in $X_i(T)$ and perturbed locations in $Z_i(T)$, where the $X_i(T)$ and $Z_i(T)$ are modeled by the first-order Markov chain and the first-order multivariate Markov chain, respectively

$$\overset{(b)}{=}\sum_{j=1}^{T}\sum_{k=j}^{T} I\left(x^{j}; z^{k} \mid x^{j-1:j-r}, z^{k-1:k-r}\right) \tag{3.9}$$

$$\overset{(c)}{=}\sum_{k=1}^{T}\sum_{j=k-r}^{k} I\left(x^{j}; z^{k} \mid x^{j-1:j-r}, z^{k-1:k-r}\right), \tag{3.10}$$

where (a) follows the fact that

$$\sum_{j=1}^{T} I\left(x^{j:1}; z^{j} \mid z^{j-1:1}\right) \tag{3.11}$$

$$=\sum_{j=1}^{T}\sum_{k=1}^{j} I\left(x^{k}; z^{j} \mid x^{k-1:1}, z^{j-1:1}\right) \tag{3.12}$$

$$=\sum_{k=1}^{T}\sum_{j=k}^{T} I\left(x^{k}; z^{j} \mid x^{k-1:1}, z^{j-1:1}\right) \tag{3.13}$$

$$=\sum_{k=1}^{T} I\left(x^{k}; z^{T:k} \mid x^{k-1:1}, z^{k-1:1}\right), \tag{3.14}$$

(b) follows the two assumptions in the definition, i.e., the real trajectory follows a Markov chain of order r and the perturbed trajectory follows a multivariate Markov chain of order r, and (c) follows the fact that $I\left(X;Y\right)=0$ if and only if the random variable X is independent of Y. Specifically, if $r=0$, i.e., the locations shared by a user are spatiotemporal uncorrelated, we have

$$\mathcal{D}(X_i(T); Z_i(T)) = I(X_i(T) \rightarrow Z_i(T)) \tag{3.15}$$

$$= \sum_{j=1}^{T} I\left(x^j; z^j\right). \tag{3.16}$$

The trajectory-level data privacy metric measures the extent of sensitive information disclosure resulting from sharing perturbed location trajectories. For instance, when the real trajectory follows a first-order Markov chain, the privacy metric quantifies the leakage of real locations x^j and x^{j-1} due to the disclosure of perturbed location z^j (as illustrated in Fig. 3.2). Thus, by minimizing this privacy metric under the given utility constraint, we can effectively protect users' location data privacy while ensuring high-quality data utility. In other words, the attacker would be unable to infer x^j and x^{j-1} from the observed perturbed location z^j.

3.3.2 Semantic Privacy Metric

As mentioned above, the attacker may deduce users' mobility intentions at real locations by analyzing the PoI distributions of sanitized locations [13, 30]. The privacy mechanism that only protects location data by altering positional coordinates remains vulnerable to semantic inference attacks when PoI characteristics of locations persist. Indeed, no matter how distortion is achieved, when two PoI distributions of the real location x^j and perturbed location z^j are similar/same, it is high-likely that the attacker can infer the purpose of the visit. For this reason, we propose the semantic privacy metric based on the mutual information to measure the leakage of the visit purpose caused by the similarity between the PoI distributions of real location and perturbed location.

According to Definition 3.1, mutual information quantifies the similarity between the PoI distributions of real and perturbed locations. Based on this, we define the semantic privacy metric as follows.

Definition 3.6 (Semantic Privacy Metric) Let $X_i(T)$ and $Z_i(T)$ represent the user u_i's real and perturbed trajectories, respectively. The semantic privacy metric for online location trajectory sharing is then defined as

$$\mathcal{S}(X_i(T); Z_i(T)) = \sum_{j=1}^{T} E\left[I\left(P(x^j); P(z^j)\right)\right], \tag{3.17}$$

where x^j is a random variable and $P(x^j)$ represents its associated PoI distribution. For each instance of real location x^j and perturbed location z^j, we calculate the mutual information $I\left(P(x^j); P(z^j)\right)$ and further take the expected mutual information with respect to the joint distribution of (x^j, z^j) to measure semantic privacy. Next, the expectation in Eq. (3.17) can be formulated as

$$E\left[I\left(P(x^j); P(z^j)\right)\right] = \sum_{x^j, z^j} p(x^j, z^j) \times I\left(P(x^j); P(z^j)\right). \tag{3.18}$$

Existing privacy-preserving mechanisms based on semantic privacy metrics do not provide provable privacy protection. For instance, Tu et al. [13] used Kullback-Leibler (KL) divergence to assess the similarity between PoI distributions at each location and across an entire city. Bindschaedler et al. [9] applied Earth Mover's Distance (EMD) to evaluate the semantic similarity between synthetic and real trajectory mobility models. Thus, these semantic privacy metrics [9, 13] are more suited for measuring semantic utility loss rather than semantic privacy leakage. In contrast, the proposed information-theoretic optimal mechanism SEITP satisfies 2ε-differential privacy [17], $2(\varepsilon+\sigma_{min})$-identifiability [37], and 4ε-mutual-information privacy [37] with $\varepsilon_d \leq \varepsilon \leq \varepsilon_d + 2\sigma_{min}$.

3.3.3 Semantic-Aware Trajectory Utility Metric

To ensure high-quality services from the crowdsensing server, the noise introduced by SEITP should be kept as moderate as possible. Numerous utility metrics have been proposed in the literature [18, 23, 38–40] to evaluate the utility of perturbed trajectories. However, these studies overlook the semantic utility of trajectories, despite their importance for delivering personalized services [31, 41]. To address this gap, we introduce a novel semantic-aware utility metric based on the Jensen-Shannon (JS) divergence.

JS divergence quantifies the distance between distributions and is used here to measure the PoI distributions of x^j and z^j, i.e., $D_{JS}(P(x^j) \mid\mid P(z^j))$. The key advantages of JS divergence are its symmetry and its always finite value, i.e.,

$$D_{JS}\left(P(x^j) \mid\mid P(z^j)\right) = D_{JS}\left(P(z^j) \mid\mid P(x^j)\right), \tag{3.19}$$

and $0 \leq D_{JS}(P(x^j) \mid\mid P(z^j)) \leq 1$. Additionally, the biggest difference from $I\left(P(x^j); P(z^j)\right)$ (semantic privacy metric) is that $D_{JS}\left(P(x^j) \mid\mid P(z^j)\right)$ (semantic utility metric) measures the information/utility loss when approximating $P(x^j)$ with $P(z^j)$, while $I\left(P(x^j); P(z^j)\right)$ measures the average amount of information about $P(x^j)$ contained in $P(z^j)$ over all possible choices of x^j and z^j. Taking the $P(x^j) = P(z^j)$ as an example, we have $I(P(x^j); P(z^j)) = H(P(x^j)) = H(P(z^j))$ and $D_{JS}(P(x^j) \mid\mid P(z^j)) = 0$. Therefore, sharing the perturbed location z^j in this case can guarantee perfect semantic utility and worse semantic privacy.

Definition 3.7 (Semantic-Aware Utility Metric) Let $X_i(T)$ and $Z_i(T)$ be the user u_i's real trajectory and perturbed trajectory, respectively. Then, the semantic-aware utility metric for online location trajectory data sharing is defined as

$$\mathcal{U}(X_i(T); Z_i(T))$$
$$= (1-\omega) \cdot \sum_{j=1}^{T} E\left[d(x^j, z^j)\right] + \omega \cdot \sum_{j=1}^{T} E\left[D_{JS}\left(P(x^j) \,||\, P(z^j)\right)\right], \tag{3.20}$$

where $d(x^j, z^j) = \|x^j - z^j\|_p$ denotes the distortion function and $\omega \in [0, 1]$ is a weight coefficient that controls the relative importance of distortion and similarity. Since x^j is a random variable and $P(x^j)$ represents its associated PoI distribution, the expectations in Eq. (3.20) can be formulated as

$$E\left[d(x^j, z^j)\right] = \sum_{\substack{k=j:j-r \\ x^{k:k-r}, z^{j:j-r}}} p(x^{k:k-r}, z^{j:j-r}) \cdot d(x^j, z^j), \tag{3.21}$$

$$E\left[D_{JS}\left(P(x^j) \,||\, P(z^j)\right)\right] = \sum_{x^j, z^j} p(x^j, z^j) \cdot D_{JS}\left(P(x^j) \,||\, P(z^j)\right). \tag{3.22}$$

Theoretically, leveraging the proposed privacy and utility metrics, the optimal privacy-preserving mechanism for online location trajectory data sharing can be designed by minimizing data and semantic privacy leakage while ensuring utility constraints. By solving the formulated multi-objective optimization problem, we derive SEITP (SEmantic-aware Information-Theoretic Privacy), which effectively balances data utility and data privacy.

3.4 Semantic-Aware Privacy Mapping Mechanism

Intuitively, stronger privacy protection leads to lower utility (i.e., reduced service quality), and vice versa. This creates a privacy-utility trade-off when designing the semantic-aware privacy-preserving location trajectory data sharing mechanism based on the proposed privacy and utility metrics (detailed in Sect. 3.4.1). Then, we solve the formulated optimization problem, finding the semantic-aware privacy mapping mechanism (detailed in Sect. 3.4.2). Through this mechanism, the server can protect users' privacy while still delivering high-quality location-based services. Lastly, Sect. 3.4.3 provides an analysis of the computational complexity involved in solving the optimization problem.

3.4.1 Constructing Optimization Model

We begin by demonstrating how the proposed metrics formulate the privacy-utility trade-off in offline trajectory data-sharing scenarios before extending the study to the online case.

Privacy-Utility Trade-Off for Offline Location Trajectory Data Sharing Scenarios A key question is determining the minimum privacy leakage under a given utility constraint from an information-theoretic perspective and designing a trajectory-level privacy protection mechanism to achieve it. This leads to the formulation of the following multi-objective non-linear optimization problem ($\mathcal{P}$), i.e.,

$$\min_{p(Z_i(T)|X_i(T))} \quad (\mathcal{D}(X_i(T); Z_i(T)), \mathcal{S}(X_i(T); Z_i(T))), \tag{3.23a}$$

$$\text{s.t.} \quad \mathcal{U}(X_i(T); Z_i(T)) \leq \eta, \tag{3.23b}$$

where η is the utility loss budget. By solving problem ($\mathcal{P}$), we derive the optimal privacy mapping mechanism $p(Z_i(T) \mid X_i(T))$. Designed for offline scenarios, this mechanism maps the real trajectory $X_i(T)$ of user u_i to the perturbed trajectory $Z_i(T)$ using $p(Z_i(T) \mid X_i(T))$, which is then shared with the semi-honest server.

Privacy-Utility Trade-Off for Online Location Trajectory Data Sharing Scenarios Based on the proposed privacy and utility metrics, we formulate the online case as the following optimization problem ($\mathcal{Q}$), i.e.,

$$\min_{\mathcal{L}} \left(\sum_{k=j-r}^{j} I\left(x^k; z^j \mid x^{k-1:k-r}, z^{j-1:j-r}\right), E\left[I\left(P(x^j); P(z^j)\right)\right] \right), \tag{3.24a}$$

$$\text{s.t.} \quad (1-\omega) \cdot E\left[d\left(x^j, z^j\right)\right] + \omega \cdot E\left[D_{JS}\left(P(x^j) \mid\mid P(z^j)\right)\right] \leq \eta_j. \tag{3.24b}$$

where the utility loss budget $\sum_{j=1}^{T} \eta_j = \eta$, $\eta_1 = \eta_2 = \cdots = \eta_T$, privacy mapping $\mathcal{L} := p(z^j \mid x^{k:k-r}, z^{j-1:j-r})$, $k \in [j-r, j]$, and $P(x^j)$ represents the PoI distribution of x^j. Similarly, we can obtain the optimal local privacy mapping $\mathcal{L}$ by solving the problem ($\mathcal{Q}$). This mechanism operates online manner, where given the user's real location sequence $x^{j:j-r}$ and perturbed location sequence $z^{j-1:j-r}$, the real location x^j is mapped to the perturbed location z^j using the local privacy mapping $\mathcal{L}_{(k=j)} := p(z^j \mid x^{j:j-r}, z^{j-1:j-r})$. The user then shares z^j with the semi-honest server at timestamp t^j.

3.4.2 Solving Optimization Model

Given that ($\mathcal{Q}$) is a high-dimensional, multi-objective nonlinear optimization problem, solving it directly is computationally infeasible. To address this challenge, we introduce a two-stage local privacy mapping mechanism that handles data privacy and semantic privacy separately.

Following the approach in [17, 42], we decompose the local privacy mapping $\mathscr{L}$ into two sequential mappings: data privacy mapping $\mathcal{M}$ and semantic privacy mapping $\mathcal{N}$. Specifically, $\mathcal{M}$ is defined as $p(y^j \mid x^{k:k-r}, y^{j-1:j-r})$, and $\mathcal{N}$ as $p(z^j \mid y^j)$. In this process, the real location x^j is first mapped to y^j through $\mathcal{M}_{(k=j)} := p(y^j \mid x^{j:j-r}, y^{j-1:j-r})$, and then y^j is further mapped to z^j using $\mathcal{N}$. Finally, the perturbed location z^j is shared with the semi-honest server at timestamp t^j to access services. By combining $\mathcal{M}_{(k=j)}$ and $\mathcal{N}$, both data privacy and semantic privacy leakage are prevented.

Building on the two-stage local privacy mapping mechanism described above, we reformulate the optimization problem ($\mathcal{Q}$) into a relaxed version ($\mathcal{R}$), i.e.,

$$\min_{\mathcal{M},\mathcal{N}} \left(\sum_{k=j-r}^{j} I\left(x^k; y^j \mid x^{k-1:k-r}, y^{j-1:j-r}\right), E\left[I\left(P(y^j); P(z^j)\right)\right]\right), \tag{3.25a}$$

$$\text{s.t.} \quad (1-\omega)\cdot E\left[d\left(x^j, y^j\right)\right] + \omega \cdot E\left[D_{JS}\left(P(y^j) \,||\, P(z^j)\right)\right] \le \eta_j. \tag{3.25b}$$

Next, we give the detailed expression of the expectation in the problem ($\mathcal{R}$), i.e.,

$$E\left[I\left(P(y^j); P(z^j)\right)\right] \overset{(d)}{=} \sum_{z^j, y^j} p(y^j)\cdot \mathcal{N}\cdot I\left(P(y^j); P(z^j)\right), \tag{3.26}$$

$$E\left[d\left(x^j, y^j\right)\right] \overset{(e)}{=} \sum_{\substack{k=j:j-r \\ x^{k:k-r}, y^{j:j-r}}} p(x^{k-1:k-r}, y^{j-1:j-r})\cdot p(x^k \mid x^{k-1:k-r}, y^{j-1:j-r})\cdot \mathcal{M}\cdot d(x^j, y^j) \tag{3.27}$$

$$E\left[D_{JS}\left(P(y^j) \,||\, P(z^j)\right)\right] \overset{(f)}{=} \sum_{z^j, y^j} p(y^j)\cdot \mathcal{N}\cdot D_{JS}\left(P(y^j) \,||\, P(z^j)\right), \tag{3.28}$$

where (d) follows the fact that Eq. (3.18) and $p(y^j) \times \mathcal{N} = p(y^j, z^j)$, (e) follows the fact that Eq. (3.21) and

$$p(x^{k-1:k-r}, y^{j-1:j-r}) \times p(x^k \mid x^{k-1:k-r}, y^{j-1:j-r}) \times \mathcal{M} = p(x^{k:k-r}, y^{j:j-r}), \tag{3.29}$$

and (f) follows the fact that Eq. (3.22) and $p(y^j) \times \mathcal{N} = p(y^j, z^j)$. By solving the optimization problem ($\mathcal{R}$), we derive the local privacy mappings $\mathcal{M}$ and $\mathcal{N}$. In the following sections, we detail the process of obtaining the stage-one local privacy mapping $\mathcal{M}$ (data privacy mapping) and the stage-two local privacy mapping $\mathcal{N}$ (semantic privacy mapping), respectively.

Stage-One Privacy Mapping In the first stage, we consider the data privacy subproblem ($\mathcal{R}_1$), i.e.,

$$\min_{\mathcal{M}} \sum_{k=j-r}^{j} I\left(x^k; y^j \mid x^{k-1:k-r}, y^{j-1:j-r}\right), \tag{3.30a}$$

$$\text{s.t. } (1-\omega) \cdot E\left[d\left(x^j, y^j\right)\right] \leq \eta_j \cdot \omega, \tag{3.30b}$$

where $d(x^j, y^j) = \|x^j - y^j\|_p$. In the following, we prove that the optimization problem ($\mathcal{R}_1$) is convex.

Theorem 3.1 *The minimization problem* ($\mathcal{R}_1$) *is a convex optimization problem.*

Proof Obviously, the feasible region of the optimization problem ($\mathcal{R}_1$) is a convex set. Therefore, we only need to prove that the objective function of ($\mathcal{R}_1$) (i.e., Right-Hand Side (RHS) of Eq. (3.30a)) is a convex function with respect to the privacy mapping $\mathcal{M}$. By definition, we have

$$I\left(x^k; y^j \mid x^{k-1:k-r}, y^{j-1:j-r}\right)$$
$$= \sum_{x^{k:k-r}, y^{j:j-r}} p(x^{k:k-r}, y^{j-1:j-r}) \cdot \mathcal{M} \cdot \log\left(\frac{\mathcal{M}}{p(y^j \mid x^{k-1:k-r}, y^{j-1:j-r})}\right). \tag{3.31}$$

We further let $f(\mathcal{M})$ denote the RHS of Eq. (3.31) and take the second derivative of $f(\mathcal{M})$, as follows

$$f''(\mathcal{M}) = \sum_{x^{k:k-r}, y^{j:j-r}} p\left(x^{k:k-r}, y^{j-1:j-r}\right) / \mathcal{M}. \tag{3.32}$$

Since $\mathcal{M} \in [0, 1]$ and $p(x^{k:k-r}, y^{j-1:j-r}) \in [0, 1]$, we derive that $f''(\mathcal{M}) \geq 0$, i.e., the objective function (i.e., RHS of Eq. (3.30a)) is a convex function with respect to the privacy mapping $\mathcal{M}$. In summary, the minimization problem ($\mathcal{R}_1$) is a convex optimization problem. □

Although the optimization problem ($\mathcal{R}_1$) is convex (as shown in Theorem 3.1), it cannot be directly solved due to the coupling of its variables, $\mathcal{M}$ and $p(y^j \mid x^{k-1:k-r}, y^{j-1:j-r})$. To address this challenge, we decouple the decision variables and solve the complex optimization problem ($\mathcal{R}_1$) using the Alternating Direction Method of Multipliers (ADMM) [43]. Specifically, we first reformulate ($\mathcal{R}_1$) into an equivalent standard form ($\widehat{\mathcal{R}}_1$) that can be efficiently handled by ADMM. Then, we apply ADMM to ($\widehat{\mathcal{R}}_1$) to derive the optimal stage-one local privacy mapping $\mathcal{M}$.

Specifically, we introduce a slack variable v to transform the inequality constraint (see Eq. (3.30b)) into an equality constraint. Additionally, we define a new variable $\widehat{\mathcal{M}}$ to reformulate the objective function (see Eq. (3.30a)) into a separable form, i.e.,

$$p(y^j \mid x^{k:k-r}, y^{j-1:j-r}) \times \log(p(y^j \mid x^{k-1:k-r}, y^{j-1:j-r})) = \widehat{\mathcal{M}}. \tag{3.33}$$

Through these transformations, the optimization problem $(\mathcal{R}_1)$ is converted into the standard form $(\widehat{\mathcal{R}}_1)$ as follows,

$$\min_{\mathcal{M},\widehat{\mathcal{M}}} \sum_{\substack{k=j:j-r \\ x^{k:k-r}, y^{j:j-r}}} p\left(x^{k:k-r}, y^{j-1:j-r}\right) \cdot \left(\mathcal{M} \cdot \log \mathcal{M} - \widehat{\mathcal{M}}\right), \tag{3.34a}$$

$$\text{s.t. } E\left[d\left(x^j, y^j\right)\right] - \frac{\omega}{1-\omega} \cdot \eta_j + v^2 = 0, \tag{3.34b}$$

$$\widehat{\mathcal{M}} - \mathcal{M} \cdot \log\left(p(y^j \mid x^{k-1:k-r}, y^{j-1:j-r})\right) = 0. \tag{3.34c}$$

Next, we present an iterative algorithm based on ADMM to solve the optimization problem $(\widehat{\mathcal{R}}_1)$. ADMM operates efficiently by alternately optimizing one variable while keeping the others fixed. To achieve this, we first derive the augmented Lagrangian function for the problem $(\widehat{\mathcal{R}}_1)$, which is given by

$$\begin{aligned} & L_\rho\left(\mathcal{M}, \widehat{\mathcal{M}}, \gamma_1, \gamma_2, v\right) \\ = & \sum_{\substack{k=j:j-r \\ x^{k:k-r}, y^{j:j-r}}} p\left(x^{k:k-r}, y^{j-1:j-r}\right) \cdot \left(\mathcal{M} \cdot log \mathcal{M} - \widehat{\mathcal{M}}\right) \\ & + \gamma_1 \cdot \left(E\left[d(x^j, y^j)\right] - \frac{\omega}{1-\omega} \cdot \eta_j + v^2\right) + \gamma_2 \cdot \widetilde{\mathcal{M}} \\ & + \frac{\rho}{2} \cdot \left((E\left[d(x^j, y^j)\right] - \frac{\omega}{1-\omega} \cdot \eta_j + v^2)^2 + \widetilde{\mathcal{M}}^2\right), \end{aligned} \tag{3.35}$$

where

$$\widetilde{\mathcal{M}} := \widehat{\mathcal{M}} - \mathcal{M} \cdot \log\left(p(y^j \mid x^{k-1:k-r}, y^{j-1:j-r})\right), \tag{3.36}$$

$\rho > 0$ is the coefficient of the quadratic penalty, and γ_1 and γ_2 are Lagrange multipliers. Next, we explain how to eliminate the slack variable v from the augmented Lagrangian function (see Eq. (3.35)). Specifically, when $\mathcal{M}$ and $\widehat{\mathcal{M}}$ are fixed, the sub-problem about v can be formulated as

$$\begin{aligned} v^* = \min_{v} \Bigg\{ & \gamma_1 \cdot \left(E\left[d(x^j, y^j)\right] - \frac{\omega}{1-\omega} \cdot \eta_j + v^2\right) \\ & + \frac{\rho}{2} \cdot \left(E\left[d(x^j, y^j)\right] - \frac{\omega}{1-\omega} \cdot \eta_j + v^2\right)^2 \Bigg\}, \end{aligned} \tag{3.37}$$

where v^* is the optimal solution function of the optimization problem Eq. (3.37). Therefore, we can eliminate the slack variable v by substituting $v = v^*$ into Eq. (3.35), so that the problem $\min L_\rho(\mathcal{M}, \widehat{\mathcal{M}}, \gamma_1, \gamma_2, v^*)$ is only about the decision variables $\mathcal{M}$ and $\widehat{\mathcal{M}}$. Finally, we give the iterative format for solving the optimization problem $(\widehat{\mathcal{R}}_1)$, i.e.,

$$\widehat{\mathcal{M}}^{k+1} = \min_{\widehat{\mathcal{M}}} L_\rho\left(\mathcal{M}^k, \widehat{\mathcal{M}}, \gamma_1^k, \gamma_2^k, v^*\right), \tag{3.38}$$

$$\mathcal{M}^{k+1} = \min_{\mathcal{M}} L_\rho\left(\mathcal{M}, \widehat{\mathcal{M}}^{k+1}, \gamma_1^k, \gamma_2^k, v^*\right), \tag{3.39}$$

$$\gamma_1^{k+1} = \gamma_1^k + \tau\rho\left(E\left[d(x^j, y^j)\right] - \frac{\omega}{1-\omega}\cdot\eta_j + (v^*)^2\right), \tag{3.40}$$

$$\gamma_2^{k+1} = \gamma_2^k + \tau\rho\cdot\widetilde{\mathcal{M}}, \tag{3.41}$$

where $\tau \in (0, (1+\sqrt{5})/2]$ is the step size.

From Eq. (3.35), it is clear that the key step in solving the minimization problem $\min L_\rho(\mathcal{M}, \widehat{\mathcal{M}}, \gamma_1, \gamma_2, v^*)$, i.e., solving Eqs. (3.38) and (3.39), is to calculate $p(x^{k:k-r}, y^{j-1:j-r})$. We expand the calculation of $p(x^{k:k-r}, y^{j-1:j-r})$ as follows,

$$p(x^{k:k-r}, y^{j-1:j-r}) \tag{3.42}$$

$$\overset{(h)}{=} p(x^{k-1:k-r}, y^{j-1:j-r})\cdot p(x^k \mid x^{k-1:k-r}) \tag{3.43}$$

$$= \sum_{\substack{x^{k-r-1}\\ y^{j-r-1}}} p(x^{k-1:k-r-1}, y^{j-1:j-r-1})\cdot p(x^k \mid x^{k-1:k-r}). \tag{3.44}$$

Equation (h) holds due to the Markov chain property of order r, where the real location x^k depends only on the preceding real location sequence $\{x^{k-1}, x^{k-2}, \cdots, x^{k-r}\}$. Therefore, we can observe that $p(x^{k:k-r}, y^{j-1:j-r})$ can be computed directly from the joint probability distribution at the previous timestamp $p(x^{k-1:k-r-1}, y^{j-1:j-r-1})$, i.e.,

$$\begin{aligned} &p(x^{k-1:k-r-1}, y^{j-1:j-r-1}) \\ =&p(y^{j-1} \mid x^{k-1:k-r-1}, y^{j-2:j-r-1})\cdot p(x^{k-1:k-r-1}, y^{j-2:j-r-1}), \end{aligned} \tag{3.45}$$

and the location transition probability $p(x^k \mid x^{k-1:k-r})$. Thus, we use the joint probability distribution $p(x^{k:k-r}, y^{j-1:j-r})$ as the input for Algorithm 1.

Once the stage-one privacy mapping $\mathcal{M}$, we can compute its data privacy leakage as follows,

$$\sum_{k=j-r}^{j} I\left(x^k; y^j \mid x^{k-1:k-r}, y^{j-1:j-r}\right) \tag{3.46}$$

$$= \sum_{\substack{k=j:j-r \\ x^{k:k-r}, y^{j:j-r}}} p\left(x^{k:k-r}, y^{j:j-r}\right) \cdot \log\left(\frac{\mathcal{M}}{p(y^j \mid y^{j-1:j-r})}\right) \tag{3.47}$$

$$= \sum_{\substack{k=j:j-r \\ x^{k:k-r}, y^{j:j-r}}} \mathcal{M} \cdot p(x^{k:k-r}, y^{j-1:j-r}) \cdot \log\left(\frac{\mathcal{M}}{p(y^j \mid y^{j-1:j-r})}\right), \tag{3.48}$$

where the transition probability $p(y^j \mid y^{j-1:j-r})$ is calculated as below,

$$p(y^j \mid y^{j-1:j-r}) = \sum_{x^{k:k-r}} \mathcal{M} \cdot \frac{p(x^{k:k-r}, y^{j-1:j-r})}{p(y^{j-1:j-r})}, \tag{3.49}$$

and the calculation method of the joint probability distribution $p(x^{k:k-r}, y^{j-1:j-r})$ is given by Eq. (3.44).

Algorithm 1 Obtain the stage-one privacy mapping

Input: $p(x^{k:k-r}, y^{j-1:j-r})$: joint probability distribution, τ: step size, ϵ: threshold for convergence of the algorithm, $d(x^{T:1}, y^{T:1})$: distortion matrix, ρ: coefficient of the quadratic penalty, $\sigma > 1$: constant;

Output: $\mathcal{M}$: stage-one privacy mapping, I_d^*: minimum value of data privacy leakage;

Initialize stage-one privacy mapping $\mathcal{M}$ and Lagrange multipliers $\boldsymbol{\gamma} = [\gamma_1, \gamma_2]$ as a uniform distribution and random value, respectively;
Calculate $\widehat{\mathcal{M}}$ using $\mathcal{M}$ and $\boldsymbol{\gamma}$ by Eq. (3.38);
Calculate I_0 using $\mathcal{M}$ and $\widehat{\mathcal{M}}$ by Eq. (3.31);
Calculate $\widehat{\mathcal{M}}$ using $\mathcal{M}$ and $\boldsymbol{\gamma}$ by Eq. (3.39);
Calculate γ_1 using $\widehat{\mathcal{M}}$ and $\mathcal{M}$ by Eq. (3.40);
Calculate γ_2 using $\widehat{\mathcal{M}}$ and $\mathcal{M}$ by Eq. (3.41);
while True **do**
  Calculate $\widehat{\mathcal{M}}$ using $\mathcal{M}$ and $\boldsymbol{\gamma}$ by Eq. (3.38);
  Calculate I using $\mathcal{M}$ and $\widehat{\mathcal{M}}$ by Eq. (3.31);
  if $|I - I_0| \le \epsilon$ **then**
    $I_d^* = I$;
    Break;
  else
    Calculate $\mathcal{M}$ using $\widehat{\mathcal{M}}$ and $\boldsymbol{\gamma}$ by Eq. (3.39);
    Calculate γ_1 using $\widehat{\mathcal{M}}$ and $\mathcal{M}$ by Eq. (3.40);
    Calculate γ_2 using $\widehat{\mathcal{M}}$ and $\mathcal{M}$ by Eq. (3.41);
    $\rho = \sigma \cdot \rho$;

$I_0 = I$;
end if
end while
Output $\mathcal{M}$ and I_d^*;

The algorithm for obtaining the optimal stage-one local privacy mapping $\mathcal{M}$ by solving the optimization problem $(\widehat{\mathcal{R}}_1)$ is provided in Algorithm 1. Since the privacy-preserving mechanism has access to previous location data, including both the real location sequence $(x^k, x^{k-1}, \cdots, x^{k-r})$ and perturbed location sequence $(y^{j-1}, y^{j-2}, \cdots, y^{j-r})$, the stage-one privacy mapping $\mathcal{M}$ is stored in a D-dimensional vector. Similarly, the decision variable $\widehat{\mathcal{M}}$ is also represented as a D-dimensional vector, where D denotes the number of spatial grids. We initialize the distribution $\mathcal{M}$ and the vector $\boldsymbol{\gamma} = [\gamma_1, \gamma_2]$ as the starting point for the iterative process. The initial stage-one privacy mapping $\mathcal{M}$ is set to a uniform distribution, i.e., $\mathcal{M} = [1/D, 1/D, \cdots, 1/D]_D$. During each iteration, we alternate between fixing the decision variable $\mathcal{M}$ and the Lagrange multipliers $\boldsymbol{\gamma}$ to minimize $\widehat{\mathcal{M}}$, followed by fixing $\widehat{\mathcal{M}}$ and $\boldsymbol{\gamma}$ to minimize $\mathcal{M}$, and finally updating the Lagrange multipliers $\boldsymbol{\gamma}$. By repeating these steps, the right-hand side of Eq. (3.30a) is minimized, thereby reducing data privacy leakage.

Stage-Two Privacy Mapping In the second stage, we consider the semantic privacy sub-problem $(\mathcal{R}_2)$, i.e.,

$$\min_{\mathcal{N}} \ E\left[I\left(P(y^j); P(z^j)\right)\right], \tag{3.50}$$

$$\text{s.t. } \omega \cdot E\left[D_{JS}\left(P(y^j) \,||\, P(z^j)\right)\right] \leq (1-\omega) \cdot \eta_j. \tag{3.51}$$

Since the PoI distribution function $P(y^j)$ is non-differentiable, gradient-based methods cannot be applied. Thus, we determine the stage-two local privacy mapping $\mathcal{N}$ by solving the optimization problem $(\mathcal{R}_2)$ using the Artificial Bee Colony (ABC) algorithm [44]. The ABC algorithm, a recent development in swarm intelligence, emulates the intelligent foraging behavior of bee colonies and operates without requiring the objective function to be differentiable. The implementation details for employing the ABC algorithm to solve $(\mathcal{R}_2)$ are outlined in Algorithm 2.

Algorithm 2 Obtain the stage-two privacy mapping

Input: SN: the size of population, D: the number of grid, MCY: the maximum number of iterations, lim: limit for non-improvement;
Output: $\mathcal{N}$: stage-two privacy mapping, I_s^*: minimum value of semantic privacy leakage;
Initialization the population solutions $\mathcal{L}_m = [L_{m1}, L_{m2}, \cdots, L_{mD}]$, $m = 1, 2, \cdots, SN$, where we have $\sum_{j=1}^{D} L_{mj} = 1$;
Calculate fitness value of the population by Eq. (3.52);
$Cycle = 1$;

4: **while** $Cycle < MCY$ **do**
5: **for** $m = 1 : SN$ **do**
6: Produce a new candidate solution $\mathcal{V}_m$ by Eqs. (3.53) and (3.54);
7: Calculate fitness value of $\mathcal{V}_m$ by Eq. (3.52);
8: Apply the selection process between $\mathcal{V}_m$ and $\mathcal{L}_m$ based on Deb's method;
9: If solution $\mathcal{L}_m$ does not improve $trial_m = trial_m + 1$, otherwise $trial_m = 0$;
10: Calculate the probability values $p(\mathcal{V}_m)$ for the solutions by Eq. (3.55);
11: **end for**
12: $t = 0, m = 1$;
13: **while** $t \leq SN$ **do**
14: **if** $random(0, 1) < p(\mathcal{V}_m)$ **then**
15: $t = t + 1$;
16: Produce a new candidate solution $\mathcal{V}_m$ by Eqs. (3.53) and (3.54);
17: Apply the selection process between $\mathcal{V}_m$ and $\mathcal{L}_m$ based on Deb's method;
18: If solution $\mathcal{L}_m$ does not improve $trial_m = trial_m + 1$, otherwise $trial_m = 0$;
19: **end if**
20: $m = m + 1$;
21: $m = m \ mod(SN + 1)$;
22: **end while**
23: **if** $max(trial_m) > lim$ **then**
24: Replace $\mathcal{L}_m$ with a new randomly produced solution;
25: **end if**
26: **end while**
27: Output optimal individual $\mathcal{N}$ and optimal value I_s^*;

During the initialization phase (lines 1–3), the ABC algorithm first creates an initial population consisting of SN randomly generated solutions, where SN denotes the population size (line 1). Each solution is represented by a D-dimensional vector $\mathcal{L}_m = [L_{m1}, L_{m2}, \cdots, L_{mD}]$ that satisfies $\sum_{j=1}^{D} L_{mj} = 1$. Next, the fitness of each solution is evaluated (line 2) by

$$fit(\mathcal{L}_m) = \frac{1}{1 + \sum_{j=1}^{D} L_{mj} \cdot I\left(P(l_m); P(l_j)\right)}, \tag{3.52}$$

where each L_{mj} represents the probability of mapping the actual location l_m to the actual location l_j.

During the employed bees phase (lines 5–11), each employed bee starts by generating a new candidate solution $\mathcal{V}_m$ using its current memory solution $\mathcal{L}_m$ and a search equation (line 6). The quality of this candidate is then assessed (line 7). In this phase, the ABC algorithm applies the following search equations,

$$X_{mj} = L_{mj} + \phi_{mj} \cdot \left(L_{mj} - L_{kj}\right), \tag{3.53}$$

$$\mathcal{V}_m = [X_{m1}, X_{m2}, \cdots, X_{mD}] / sum\left(\mathcal{X}_m\right), \tag{3.54}$$

where m and k are distinct indices selected randomly from the range $[1, SN]$, ϕ_{mj} is a random number in $[-1, 1]$, and $\mathcal{X}_m = [X_{m1}, X_{m2}, \cdots, X_{mD}]$. After generating the candidate solution, the algorithm enters the selection stage (line 8). To satisfy the constraints of problem $(\mathcal{R}_2)$, Algorithm 2 incorporates Deb's rules [45] rather than a simple greedy approach. Deb's method employs a tournament selection operator that compares two solutions at a time using the following criteria:

- Feasibility Preference: A solution that meets all constraints (i.e., $vio(\mathcal{L}_m) \leq 0$) is always chosen over one that violates any constraint (i.e., $vio(\mathcal{L}_m) > 0$);
- Fitness Comparison Among Feasible Solutions: Between two feasible solutions (both with $vio(\mathcal{L}_m) \leq 0$ and $vio(\mathcal{L}_j) \leq 0$), the one with the better fitness value is selected (i.e., if $fit(\mathcal{L}_m) < fit(\mathcal{L}_j)$, then $\mathcal{L}_j$ is deemed dominant);
- Constraint Violation Comparison Among Infeasible Solutions: When comparing two infeasible solutions (both with $vio(\mathcal{L}_m) > 0$ and $vio(\mathcal{L}_j) > 0$), the solution with the smaller constraint violation is favored (i.e., if $vio(\mathcal{L}_m) < vio(\mathcal{L}_j)$, then $\mathcal{L}_i$ is dominant).

Once all employed bees have finished their search, they share their solutions (i.e., nectar information) with the onlooker bees in the dance area by computing probability values (line 10). These probabilities are calculated as follows,

$$p(\mathcal{V}_m) = \begin{cases} \dfrac{fit(\mathcal{V}_m)}{\sum_{j=1}^{SN} fit(\mathcal{V}_j)} \times 0.5 + 0.5 \ if \ vio(\mathcal{V}_m) \leq 0, \\ (1 - \dfrac{vio(\mathcal{V}_m)}{\sum_{j=1}^{SN} vio(\mathcal{V}_j)}) \times 0.5 \ if \ vio(\mathcal{V}_m) > 0, \end{cases} \tag{3.55}$$

with the constraint violation function defined by

$$vio(\mathcal{L}_m) = \sum_{j=1}^{D} L_{mj} \cdot D_{JS}\left(P(l_m) \mid\mid P(l_j)\right) - \frac{1-\omega}{\omega} \cdot \eta_j. \tag{3.56}$$

During the onlooker bees phase (lines 13–22), each onlooker bee selects a candidate solution based on its computed probability (line 14) and then makes modifications to that solution (line 16). The revised solution's quality is subsequently assessed by calculating its fitness (line 17), in a manner similar to the employed bees.

During the scout bees phase (lines 23–25), the algorithm identifies and discards nectar sources (i.e., solutions) that have not shown improvement over a predetermined number of cycles (denoted as lim). These underperforming solutions are then replaced with new candidate solutions discovered by the scout bees (line 24).

Ultimately, the process yields the optimal individual $\mathcal{N}$ along with the optimal value I_s^*, which correspond to the best stage-two privacy mapping and the minimal semantic privacy leakage, respectively.

3.4.3 Computational Complexity

In this section, we evaluate the computational complexity of SEITP. First, we solve problem ($\mathcal{R}_1$) using Algorithm 1 and problem ($\mathcal{R}_2$) using Algorithm 2 to derive the stage-one privacy mapping $\mathcal{M}$ and stage-two privacy mapping $\mathcal{N}$, respectively. These mappings are then used to implement SEITP. Accordingly, the overall computational complexity of SEITP comprises that of Algorithm 1 (ADMM) and Algorithm 2 (ABC algorithm), which we now analyze using Big O notation.

Algorithm 1 involves optimizing two decision variables: the auxiliary variable $\widehat{\mathcal{M}}$ and the stage-one privacy mapping $\mathcal{M}$. Since the algorithm alternates between optimizing $\mathcal{M}$ (as detailed in Eq. (3.39)) and $\widehat{\mathcal{M}}$ (as detailed in Eq. (3.38)), each operation requires $O(D^3)$, resulting in a combined complexity of $O(D^3) + O(D^3)$.

In contrast, Algorithm 2 focuses on a single decision variable, namely the stage-two privacy mapping $\mathcal{N}$. Here, we concentrate on the computational cost during the employed bees phase (step 2) and the onlooker bees phase (step 3). Within each iteration, the employed bees phase requires $O(2 \cdot SN \cdot D^2)$, while the onlooker bees phase takes $O(SN \cdot D^2)$. Consequently, the total complexity for Algorithm 2 is $O(2 \cdot SN \cdot D^2) + O(SN \cdot D^2)$.

By combining these results, the overall computational complexity of SEITP becomes $O(D^3 + SN \cdot D^2)$. This complexity is simplified to $O(D^3)$ when $D \gg SN$ and to $O(SN \cdot D^2)$ when $SN \gg D$. Although the computational cost is relatively high, the two-stage privacy mapping ($\mathcal{M}$ and $\mathcal{N}$) can be precomputed offline, such as during periods of low usage or at night, enabling efficient sharing of the perturbed location once the mappings are established.

3.5 Privacy Analysis

This section begins by introducing the definitions of identifiability [37], differential privacy [17], and mutual-information privacy [37]. Then, leveraging the completeness privacy notions framework detailed in Chap. 2, we demonstrate that SEITP effectively protects both data and semantic privacy. Specifically, SEITP meets the privacy guarantees of 2ε-DP, $2(\varepsilon + \sigma_{min})$-identifiability, and 4ε-mutual-information privacy with $\varepsilon_d \leq \varepsilon \leq \varepsilon_d + 2\sigma_{min}$.

Definition 3.8 (Identifiability) A mechanism $\mathcal{A}$ satisfies ε_i-identifiability for some $\varepsilon_i \in \mathbb{R}^+$ if for any pair of neighboring real locations $x, x' \in \mathcal{D}^n$ and any perturbed location $z \in \mathcal{D}^n$,

$$p_{X|Z}(x \mid z) \leq e^{\varepsilon_i} \cdot p_{X|Z}(x' \mid z). \tag{3.57}$$

Definition 3.9 (Differential Privacy) A mechanism $\mathcal{A}$ satisfies ε_d-differential privacy for some $\varepsilon_d \in \mathbb{R}^+$ if for any pair of neighboring real locations $x, x' \in \mathcal{D}^n$ and any output $z \in \mathcal{D}^n$,

$$p_{Z|X}(z \mid x) \leq e^{\varepsilon_d} \cdot p_{Z|X}(z \mid x'). \tag{3.58}$$

Definition 3.10 (Mutual-Information Privacy) A mechanism $\mathcal{A}$ satisfies ε_m-mutual-information privacy for some $\varepsilon_m \in \mathbb{R}^+$ if the mutual information between X and Z satisfies $I(X; Z) \leq \varepsilon_m$, where

$$I(X; Z) = \sum_{x,z \in \mathcal{D}^n} p_{X,Z}(x, z) \cdot \log \frac{p_{X,Z}(x, z)}{p_X(x) \cdot p_Z(z)}. \tag{3.59}$$

Identifiability is characterized by the inability to distinguish between any two neighboring locations from a Bayesian perspective. In essence, identifiability ensures that the posterior probabilities remain indistinguishable.

In Chap. 2, we proposed a completeness framework that interrelates identifiability, differential privacy, and mutual-information privacy. Additionally, previous studies [37] have examined the connections among optimal privacy mapping mechanisms based on these privacy notions. Three main conclusions emerge: First, Wang et al. [37] demonstrated that the mechanism minimizing mutual-information privacy leakage under a utility constraint, referred to as the mutual-information optimal mechanism, satisfies ε-DP with $\varepsilon_d \leq \varepsilon \leq \varepsilon_d + 2\sigma_{min}$. Here, σ_{min} (detailed in Definition 2.8) is a constant derived from the prior distribution p_X of the real trajectory X and is formulated as

$$\sigma_{min} = \min_{x,x' \in \mathcal{D}^n : x \sim x'} \log \frac{p_X(x)}{p_X(x')}. \tag{3.60}$$

It is important to note that if X follows a uniform distribution, then $\sigma_{min} = 0$, meaning that $\varepsilon = \varepsilon_d$. Second, Theorem 2.1 and Theorem 2.2 establish a two-way relationship between identifiability and differential privacy: ε_i-identifiability implies $(\varepsilon_i - \sigma_{min})$-DP, and conversely, ε_d-DP implies $(\varepsilon_d + \sigma_{min})$-identifiability. Third, Theorem 2.5 shows that any mechanism satisfying ε_d-DP will inherently provide $2\varepsilon_d$-mutual-information privacy. Consequently, by proving that the optimal local privacy mappings $\mathcal{M}$ and $\mathcal{N}$ satisfy ε-DP with $\varepsilon_d \leq \varepsilon \leq \varepsilon_d + 2\sigma_{min}$, we conclude that SEITP meets the privacy guarantees of 2ε-DP, $2(\varepsilon + \sigma_{min})$-identifiability, and 4ε-mutual-information privacy. The detailed theorems and proofs are provided as follows.

Theorem 3.2 *The optimal data privacy mapping $\mathcal{M}$ satisfies ε-differential privacy with $\varepsilon_d \leq \varepsilon \leq \varepsilon_d + 2\sigma_{min}$.*

Proof For two locations x^j and z^j, the mutual information is given by $I\left(x^j; z^j\right)$. Based on the definition of trajectory-level data privacy metric (the objective of sub-problem $(\mathcal{R}_1)$), we obtain

$$\sum_{k=j-r}^{j} I\left(x^k; z^j \mid x^{k-1:k-r}, z^{j-1:j-r}\right) \tag{3.61}$$

$$\leq (r+1) \cdot \max_k I\left(x^k; z^j \mid x^{k-1:k-r}, z^{j-1:j-r}\right) \tag{3.62}$$

$$\stackrel{(i)}{\leq} (r+1) \cdot \max_k I\left(x^k; z^j \mid z^{j-1}\right), \quad k \in [j-r, j] \tag{3.63}$$

$$= (r+1) \cdot I(x^j; z^j \mid z^{j-1}) \tag{3.64}$$

$$\leq (r+1) \cdot I(x^j; z^j), \tag{3.65}$$

where (i) follows the fact that conditioning does not increase mutual information, i.e., $I(X; Y \mid Z) = I(X; Y) - I(X; Y; Z)$. We denote the standard mutual-information optimization problem as $(\mathcal{O})$, i.e.,

$$\min_{p(z^j|x^j)} \quad I(x^j; z^j) \tag{3.66a}$$

$$\text{s.t.} \quad (1-\omega) \cdot E[d(x^j, y^j)] \leq \eta_j \cdot \omega. \tag{3.66b}$$

Then, we give the optimization problem $(\mathcal{O}_1)$ below, i.e., minimizing Eq. (3.65) given utility constrain,

$$\min_{p(z^j|x^j)} \quad (r+1) \cdot I(x^j; z^j) \tag{3.67a}$$

$$\text{s.t.} \quad (1-\omega) \cdot E[d(x^j, y^j)] \leq \eta_j \cdot \omega. \tag{3.67b}$$

We relax the problem $(\mathcal{O}_1)$ into the sub-problem $(\mathcal{R}_1)$ as a result of Eqs. (3.61) to (3.65), i.e.,

$$\sum_{k=j-r}^{j} I\left(x^k; z^j \mid x^{k-1:k-r}, z^{j-1:j-r}\right) \leq (r+1) \cdot I(x^j; z^j). \tag{3.68}$$

Since relaxation enlarges the feasible set, any optimal solution of $(\mathcal{O}_1)$ is also a solution (either optimal or sub-optimal) for $(\mathcal{R}_1)$. Moreover, Wang et al. [37] demonstrated that the optimal solution of $(\mathcal{O})$ satisfies ε-DP with $\varepsilon_d \leq \varepsilon \leq \varepsilon_d + 2\sigma_{min}$. Because the only difference between the objective functions of $(\mathcal{O})$ and $(\mathcal{O}_1)$ is the factor $r+1$, the optimal solution for $(\mathcal{O}_1)$ is identical to that of $(\mathcal{O})$. Thus, the suboptimal solution of $(\mathcal{R}_1)$, which corresponds to the optimal solutions of both $(\mathcal{O}_1)$ and $(\mathcal{O})$, satisfies ε-DP. Consequently, the data privacy mapping $\mathcal{M}$ (i.e., the optimal solution of $(\mathcal{R}_1)$) meets ε-DP with $\varepsilon_d \leq \varepsilon \leq \varepsilon_d + 2\sigma_{min}$. □

Theorem 3.3 *The optimal semantic privacy mapping $\mathcal{N}$ satisfies ε-differential privacy with $\varepsilon_d \le \varepsilon \le \varepsilon_d + 2\sigma_{min}$.*

Proof For two PoI distributions $P(x^j)$ and $P(z^j)$, the mutual information between $P(x^j)$ and $P(z^j)$ can be formulated as $I(P(x^j); P(z^j))$. Let $\widehat{I}(P(l_i); P(l_j)) := \max_{x^j, z^j} I(P(x^j); P(z^j))$, where l_i and l_j are real locations. According to the definition of semantic privacy metric (the objective of semantic privacy sub-problem $(\mathcal{R}_2)$), we have

$$E\left[I\left(P(x^j); P(z^j)\right)\right] \tag{3.69}$$

$$=\sum_{x^j, z^j} p\left(x^j\right) \times p\left(z^j \mid x^j\right) \times I\left(P(x^j); P(z^j)\right) \tag{3.70}$$

$$\le \sum_{x^j, z^j} p\left(x^j\right) \times p\left(z^j \mid x^j\right) \times \widehat{I}\left(P(l_i); P(l_j)\right) \tag{3.71}$$

$$=\widehat{I}\left(P(l_i); P(l_j)\right). \tag{3.72}$$

We denote the standard mutual-information optimization problem $(\mathcal{O})$, i.e.,

$$\min_{p(l_j|l_i)} \quad \widehat{I}(P(l_i); P(l_j)) \tag{3.73a}$$

$$\text{s.t.} \quad \omega \cdot E[D_{JS}(P(l_i), P(l_j)] \le \eta_j \cdot (1-\omega). \tag{3.73b}$$

We relax the problem $(\mathcal{O})$ into sub-problem $(\mathcal{R}_2)$, since $E\left[I\left(P(x^j); P(z^j)\right)\right] \le \widehat{I}\left(P(l_i); P(l_j)\right)$ (see Eqs. (3.69) to (3.72)). Similar to Theorem 3.2, we have the optimal semantic privacy mapping $\mathcal{N}$ (i.e., the optimal solution of problem $(\mathcal{R}_2)$) satisfies ε-DP with $\varepsilon_d \le \varepsilon \le \varepsilon_d + 2\sigma_{min}$. □

3.6 Performance Evaluation

In this section, we evaluate the effectiveness of SEITP in protecting data privacy and semantic privacy while ensuring utility on a real-world private vehicle trajectory dataset.

3.6.1 Simulation Settings

All experiments are conducted using the same dataset and implementation environment. The simulations are implemented in Python and performed on a 64-bit server with 14-core, 28-thread, 2.0GHz CPU, and 128G RAM.

Datasets We use the real-world private vehicle trajectory dataset [46] for experimental evaluation. This dataset comprises trip records from 5109 unique vehicles over the period from August 31, 2018, to September 30, 2018, totaling 329,035 trips. Each record includes the vehicle ID, departure location, departure timestamp, arrival location, and arrival timestamp. Table 3.1 presents two sample trip records from a vehicle with ID 92221. Additionally, we employ a PoI dataset collected from a mapping service, which contains 100,690 PoIs spanning 24 categories (such as Food, Shopping, and Education). Each PoI record provides details on its location, category, and other relevant information, as illustrated by two sample records in Table 3.2.

Baseline We benchmark SEITP against five related works: Geo-I [16], MCEAL (Maximizing the Conditional Entropy (i.e., privacy metric) given Average Loss (i.e., quality loss metric)) [40], OLTP [23], PTSA [13], and MK-LH [26]. Geo-I and MCEAL are designed for sporadic location sharing scenarios and do not take into account the semantic characteristics of trajectories. Consequently, they are unable to mitigate data privacy leakage arising from spatiotemporal correlations, prevent semantic privacy breaches, and maintain the semantic utility of the perturbed trajectories. In contrast, OLTP employs mutual information to quantify data privacy leakage and proposes an optimal privacy mechanism by minimizing the mutual information between the perturbed and real locations under a utility loss constraint. Similarly, MK-LH incorporates spatiotemporal correlations via a Markov model to develop an optimal location hiding strategy. However, neither OLTP nor MK-LH addresses semantic privacy leakage and ensures semantic utility. PTSA adopts a greedy algorithm to merge trajectories optimally, ensuring that the merged data meets the criteria for k-anonymity, l-diversity, and t-closeness, thereby offering resistance to semantic attacks. The limitation of PTSA is that it operates exclusively in offline scenarios, as it requires the entire trajectory to be available, which is impractical for online applications. Although another mechanism that incorporates semantic features was proposed in [40], it only applies to sporadic location sharing. For a fair comparison in trajectory sharing scenarios, we have chosen PTSA to evaluate semantic privacy leakage.

Metrics We evaluate performance using the following key metrics. Privacy leakage is assessed from two perspectives. First, to accurately quantify data privacy leakage caused by spatiotemporal correlations within a trajectory and the causal relationship between the real and perturbed trajectories, we model user mobility as an order-r Markov chain and introduce a trajectory-level data privacy metric (see Definition 3.5). Second, to measure semantic privacy leakage, we propose a semantic privacy metric based on information-theoretic privacy (see Definition 3.6). To demonstrate that SEIPT effectively protects both data and semantic privacy, we theoretically prove that the information-theoretic optimal mechanism (i.e., SEIPT) designed using these metrics satisfies ε-differential privacy (see Theorem 3.2 and Theorem 3.3). Accordingly, we employ the trajectory-level data privacy metric and the semantic privacy metric to assess data and semantic privacy leakage,

Table 3.1 Sample trip records in the private vehicle trajectory dataset

ObjectID	StartTime	StartLon	StartLat	StopTime	StopLon	StopLat
92221	12/09/18 05:09:37	114.030824	22.666193	12/09/18 05:19:36	114.008863	22.686628
92221	12/09/18 05:24:53	113.937844	22.672121	12/09/18 05:27:53	113.899174	22.659251

Table 3.2 Sample PoI records in the PoI dataset

LocationLon	LocationLat	Category	Detailed information
113.920569	22.489742	Food	Chinese restaurant
113.895077	22.485029	Shopping	Home improvement

respectively. Additionally, we evaluate the utility of the perturbed trajectory dataset through geo-data analysis tasks, including: (1) Location Service: Users share their location online (e.g., via Google Maps) to receive services such as personalized navigation. We compute the semantic-aware utility metric (see Definition 3.7) to compare the perturbed and real trajectories. (2) Top-20 PoI Extraction: Identifying the 20 most frequently visited locations. We measure the accuracy of visit rankings using precision@20.

Parameter Setting As described in the previous discussion of SEITP, we specify the parameter settings used in our experiments. For Algorithm 1 (which derives the stage-one local privacy mapping $\mathcal{M}$), we set the utility loss budget to $\eta = 1.5$ and the weight coefficient to $\omega = 0.3$. The term $\frac{\omega}{1-\omega} \cdot \eta$ determines the data utility loss constraint assigned to trajectory $U_i(T)$. Additionally, following [47], we set the step size to $\tau = 1$, the convergence threshold to $\varepsilon = 10^{-3}$, the quadratic penalty coefficient to $\rho = 0.5$, and the constant to $\sigma = 1.1$. These parameters collectively influence the effectiveness of the ADMM algorithm. For Algorithm 2 (which derives the stage-two local privacy mapping $\mathcal{N}$), we set the search radius for Points of Interest (PoIs) around locations to $r = 1.0\,km$. This radius plays a critical role in defining the PoI distribution function $P(*)$ (see Eqs. (3.50) and (3.51)), and the term $\frac{1-\omega}{\omega} \cdot \eta$ determines the semantic utility loss constraint for trajectory $U_i(T)$. Furthermore, following [44], we configure the Artificial Bee Colony (ABC) algorithm with a population size of $SN = 100$, a maximum iteration count of $MCY = 1000$, and a non-improvement limit of $lim = 20$. These parameters collectively impact the optimization performance of the ABC algorithm. It is important to note that the values of ω, η and r are determined based on empirical experiments, with detailed results presented in Sect. 3.6.2.

3.6.2 Simulation Results

In this section, we assess the privacy preservation effectiveness of SEITP using the previously defined metrics, comparing its performance against baseline models.

SEITP Prevents Data Privacy Leakage For the private vehicle trajectory dataset, as shown in Fig. 3.3a, SEITP, OLTP, and MK-LH demonstrate superior performance over Geo-I and MCEAL. Given the same utility loss, their trajectory-level data privacy leakage is significantly lower than that of Geo-I and MCEAL. This advantage stems from the fact that Geo-I and MCEAL do not account for data

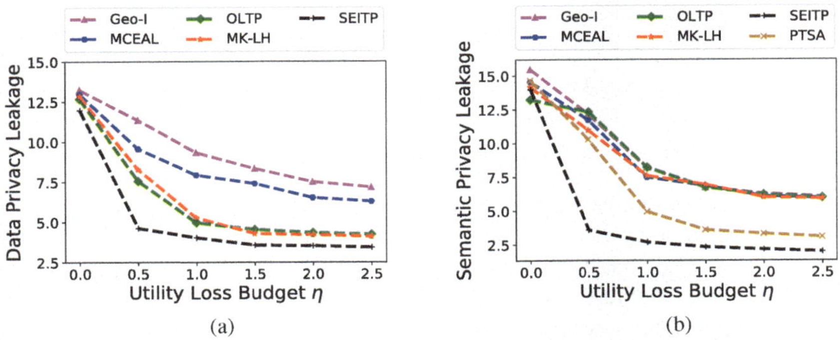

Fig. 3.3 Privacy leakage of different location-privacy preserving mechanism (SEITP and compared methods). (**a**) Data privacy leakage. (**b**) Semantic privacy leakage

privacy leakage caused by spatiotemporal correlations among locations within a trajectory. Moreover, SEITP offers stronger privacy protection compared to OLTP and MK-LH. This is because OLTP's data privacy metric only considers the current and one-step past locations, while MK-LH overlooks data privacy leakage resulting from indirect causal correlations. However, as shown in Fig. 3.2, spatiotemporal correlations, direct causal correlations, and indirect causal correlations all contribute to data privacy leakage. Additionally, we examine the impact of the utility loss budget η, the weight coefficient ω, and the radius of the PoI search area r on the data privacy, as measured by the proposed trajectory-level data privacy metric (see Fig. 3.4, black bar). The results indicate that as η and ω increase, the trajectory-level data privacy metric decreases. This is because lower values of η and ω correspond to better data utility. In contrast, the radius r primarily influences semantic utility and semantic privacy (see semantic privacy sub-problem ($\mathcal{R}_2$)), leaving the trajectory-level data privacy metric largely unaffected. Furthermore, Fig. 3.4 (black bar vs. red bar) reveals that the stage-two privacy protection does not significantly impact data privacy. In summary, the second stage of privacy protection does not substantially compromise the effectiveness of the first stage.

SEITP Prevents Semantic Privacy Leakage Figure 3.3b illustrates the trade-off between semantic privacy and utility for SEITP and the baseline methods on the private vehicle trajectory dataset. The results indicate that SEITP significantly outperforms Geo-I, MCEAL, OLTP, and MK-LH in terms of semantic privacy. This advantage arises because these methods fail to account for semantic privacy leakage caused by the PoI distribution at various locations. Compared to PTSA, SEITP offers only a moderate improvement in semantic privacy protection, as both approaches integrate PoI distribution-based semantic privacy leakage into their design. However, PTSA is limited to offline location trajectory data release, requiring full trajectory observation. Additionally, PTSA mandates that published trajectories align with the PoI distribution of an entire city rather than a localized

Fig. 3.4 Privacy leakage in the private vehicle trajectory dataset varying with the utility loss budget η, the weight coefficient ω and the radius of the circle searching PoIs r. (**a**) Privacy leakage (measured by the trajectory-level data privacy metric or semantic privacy metric) varying with η. (**b**) Privacy leakage varying with ω. (**c**) Privacy leakage varying with r

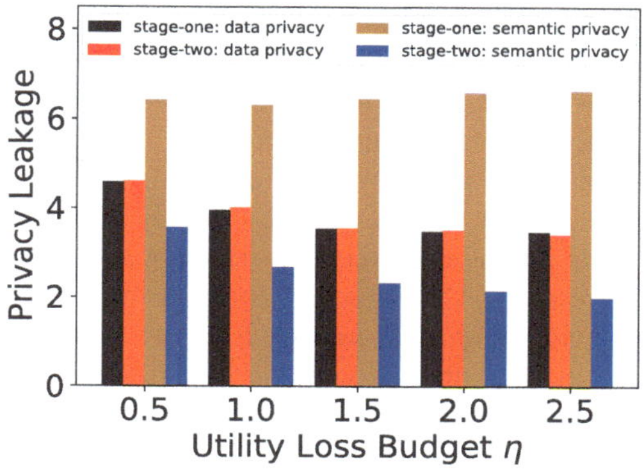

(a) Impact of η on privacy

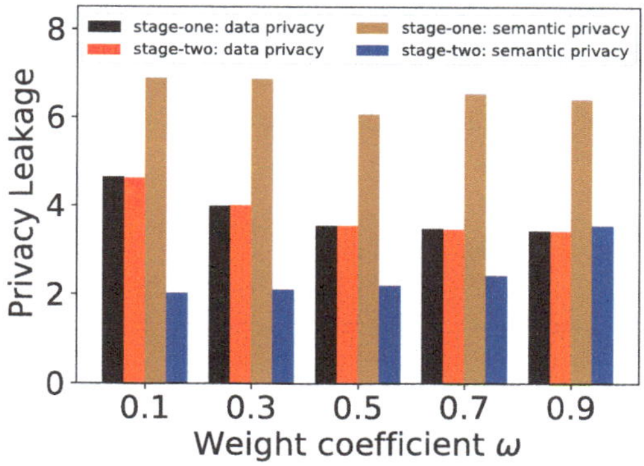

(b) Impact of ω on privacy

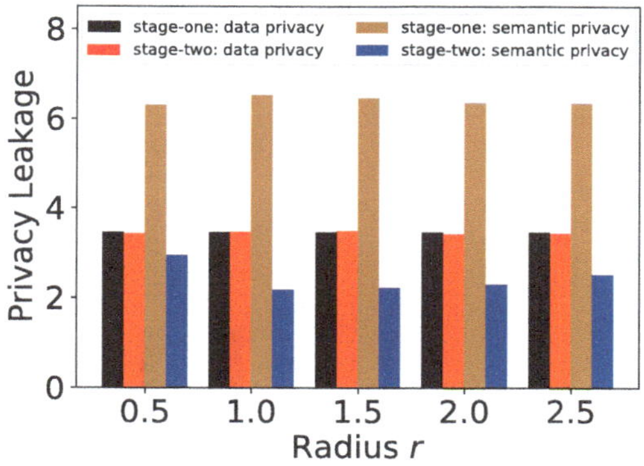

(c) Impact of r on privacy

region, which is a stringent requirement that significantly impacts data utility. Furthermore, as shown in Fig. 3.4 (blue bar), we analyze how variations in η, ω, and r affect semantic privacy. The results reveal that the semantic privacy metric decreases as η increases and rises as ω increases. This is because a larger η and a smaller ω reduce semantic utility. The radius r also plays a crucial role: if r is too small, it fails to capture a sufficient number of PoI categories, whereas if it is too large, the retrieved PoIs may not accurately reflect the semantic characteristics of locations. Consequently, the semantic privacy metric initially decreases and then increases with r. Moreover, Fig. 3.4 (tawny bar vs. blue bar) highlights that stage-one privacy protection alone is ineffective against semantic attacks, as the tawny bars remain largely unchanged across different values of η, ω, and r. This finding confirms that protecting data privacy alone is insufficient to prevent semantic privacy leakage. In summary, SEITP outperforms all existing methods in preventing both data privacy and semantic privacy leakage.

Trajectory Data Utility Guarantee We quantify SEITP's performance in geo-data analytical tasks through three parameters η, ω, and r using established evaluation metrics. Figure 3.5 demonstrates an inverse correlation between utility preservation

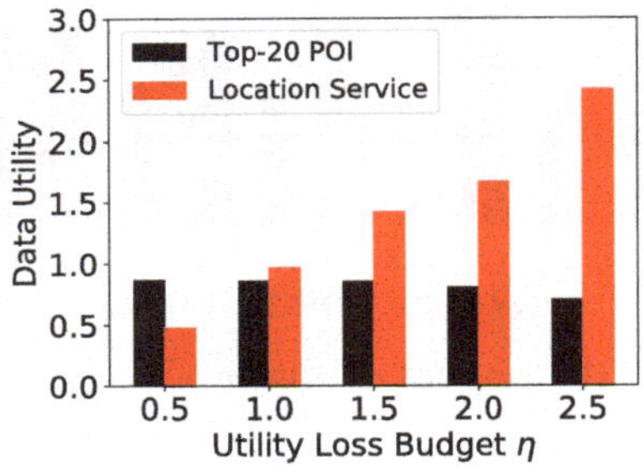

(a) Impact of η on utility

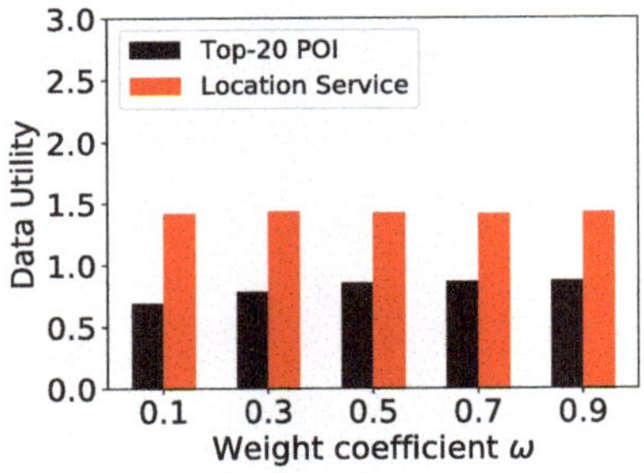

(b) Impact of ω on utility

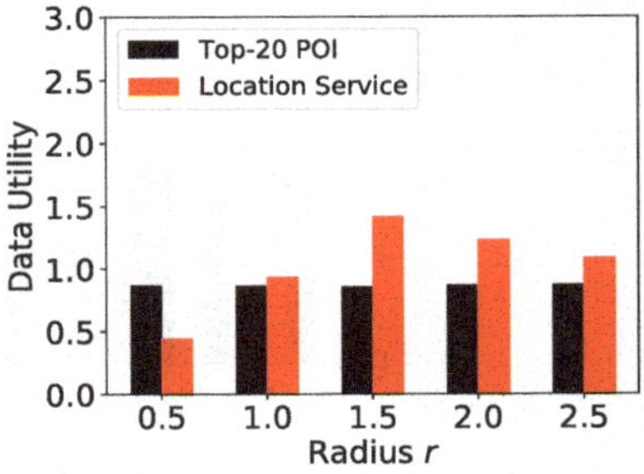

(c) Impact of r on utility

Fig. 3.5 Data utility in the private vehicle trajectory dataset varying with the utility loss budget η, the weight coefficient ω and the radius of the circle searching PoIs r. (**a**) Data utility varying with η. (**b**) Data utility varying with ω. (**c**) Data utility varying with r

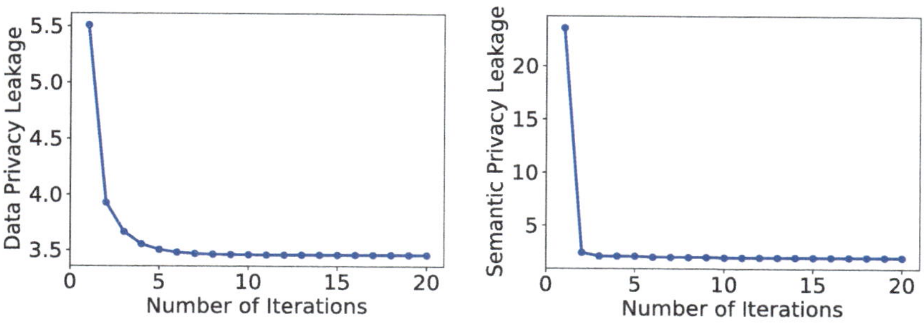

(a) Convergence of Algorithm 1 (b) Convergence of Algorithm 2

Fig. 3.6 Convergence of the proposed algorithms in the private vehicle trajectory dataset. (**a**) Convergence of the Algorithm 1 for $\tau = 1$, $\varepsilon = 10^{-3}$, $\rho = 0.5$ and $\sigma = 1.1$. (**b**) Convergence of the Algorithm 2 for $SN = 100$, $MCY = 1000$ and $lim = 20$

and data utility budget η. Specifically, semantic-aware utility metric escalates while precision@20 declines as η increases, confirming that reduced η values enhance practical usability. The weight coefficient ω primarily modulates the proportion between semantic utility and total utility allocation (Eq. 3.20), resulting in stable semantic utility measurements coupled with precision@20 improvements under elevated ω conditions. Additionally, when r is too small, it becomes challenging to retrieve a sufficient number or variety of PoI categories. Conversely, if r is too large, the retrieved PoIs may not accurately reflect the semantic characteristics of the locations. As a result, an excessively large or small radius compromises the semantic utility of the distorted trajectory. This leads to the proposed precision@20 remaining stable, while the semantic-aware utility metric initially increases and then declines.

Convergence Analysis In this part, we analyze the convergence of the proposed algorithm (Algorithms 1 and 2), specifically the number of iterations required for them to stabilize. Figure 3.6b presents the convergence curve of Algorithm 1 under the parameter settings $\tau = 1$, $\varepsilon = 10^{-3}$, $\rho = 0.5$ and $\sigma = 1.1$. The results indicate that convergence is generally achieved within approximately five iterations. Similarly, Fig. 3.6b illustrates the convergence behavior of Algorithm 2 with $SN = 100$, $MCY = 1000$, and $lim = 20$. The algorithm typically converges within about three iterations. In conclusion, the proposed algorithms reach optimal or suboptimal solutions in only a few iterations, ensuring that the computational overhead remains minimal. As a result, the convergence speed does not hinder the practical deployment of this solution in online applications.

3.7 Summary and Further Reading

In this chapter, we explored online location trajectory data sharing under a semi-honest crowdsensing server and proposed SEITP, a semantic-aware privacy-preserving mechanism. To balance privacy protection (including both data and semantic privacy) and utility (i.e., semantic-aware data utility), we formulated the problem as a multi-objective nonlinear optimization problem. Given the complexity of directly solving this optimization problem, we decomposed it into two sub-problems, i.e., data privacy preservation and semantic privacy preservation. SEITP employs a two-stage privacy protection approach, solving each sub-problem separately to protect data and semantic privacy. Theoretical analysis and extensive experiments using a real-world private vehicle trajectory dataset confirmed SEITP's effectiveness in preventing both data and semantic privacy leaks. Additionally, experimental results demonstrated SEITP's capability to preserve the semantic characteristics of trajectory data.

References

1. Jiang, H., Li, J., Zhao, P., Zeng, F., Xiao, Z., Iyengar, A.: Location privacy-preserving mechanisms in location-based services: a comprehensive survey. ACM Comput. Surv. **54**(1), 1–36 (2021)
2. Chen, L., Mislove, A., Wilson, C.: Peeking beneath the hood of uber. In: Proceedings of ACM Internet Measurement Conference, pp. 495–508 (2015)
3. Xu, F., Zhang, P., Li, Y.: Context-aware real-time population estimation for metropolis. In: Proceedings of ACM International Joint Conference on Pervasive and Ubiquitous Computing, pp. 1064–1075 (2016)
4. Ding, D., Zhang, M., Pan, X., Wu, D., Pu, P.: Geographical feature extraction for entities in location-based social networks. In: Proceedings of World Wide Web Conference, pp. 833–842 (2018)
5. Li, Z., Zheng, Z., Guo, S., Guo, B., Xiao, F., Ren, K.: Disguised as privacy: data poisoning attacks against differentially private crowdsensing systems. IEEE Trans. Mobile Comput. **22**(9), 5155–5169 (2023)
6. de Montjoye, Y.A., Hidalgo, C.A., Verleysen, M., Blondel, V.D.: Unique in the crowd: the privacy bounds of human mobility. Sci. Rep. **3**(1), 2045–2322 (2013)
7. Xu, T., Cai, Y.: Location anonymity in continuous location-based services. In: Proceedings of ACM International Symposium on Advances in Geographic Information Systems, pp. 1–8 (2007)
8. Deldar, F., Abadi, M.: PLDP-TD: personalized-location differentially private data analysis on trajectory databases. Pervasive Mobile Comput. **49**, 1–22 (2018)
9. Bindschaedler, V., Shokri, R.: Synthesizing plausible privacy-preserving location traces. In: Proceedings of IEEE Symposium on Security and Privacy, pp. 546–563 (2016)
10. Zhao, P., Jiang, H., Li, J., Zeng, F., Zhu, X., Xie, K., Zhang, G.: Synthesizing privacy preserving traces: enhancing plausibility with social networks. IEEE/ACM Trans. Netw. **27**(6), 2391–2404 (2019)
11. He, X., Cormode, G., Machanavajjhala, A., Procopiuc, C.M., Srivastava, D.: DPT: differentially private trajectory synthesis using hierarchical reference systems. Proc. VLDB Endow. **8**(11), 1154–1165 (2015)

12. Ouyang, K., Shokri, R., Rosenblum, D.S., Yang, W.: A non-parametric generative model for human trajectories. In: Proceedings of International Joint Conference on Artificial Intelligence, pp. 3812–3817 (2018)
13. Tu, Z., Zhao, K., Xu, F., Li, Y., Su, L., Jin, D.: Beyond k-anonymity: protect your trajectory from semantic attack. In: Proceedings of IEEE International Conference on Sensing, Communication, and Networking, pp. 1–9 (2017)
14. Gursoy, M.E., Liu, L., Truex, S., Yu, L.: Differentially private and utility preserving publication of trajectory data. IEEE Trans. Mobile Comput. **18**(10), 2315–2329 (2019)
15. Zheng, Z., Li, Z., Li, J., Jiang, H., Li, T., Guo, B.: Utility-aware and privacy-preserving trajectory synthesis model that resists social relationship privacy attacks. ACM Trans. Intell. Syst. Technol. **13**(3), 1–28 (2022)
16. Andrés, M.E., Bordenabe, N.E., Chatzikokolakis, K., Palamidessi, C.: Geo-indistinguishability: differential privacy for location-based systems. In: Proceedings of ACM SIGSAC Conference on Computer and Communications Security, pp. 901–914 (2013)
17. Sun, M., Tay, W.P.: On the relationship between inference and data privacy in decentralized iot networks. IEEE Trans. Inf. Foren. Secur. **15**, 852–866 (2020)
18. Xiao, Y., Xiong, L.: Protecting locations with differential privacy under temporal correlations. In: Proceedings of ACM SIGSAC Conference on Computer and Communications Security, pp. 1298–1309 (2015)
19. Jiang, H., Zhao, P., Wang, C.: Roblop: towards robust privacy preserving against location dependent attacks in continuous LBS queries. IEEE/ACM Trans. Netw. **26**(2), 1018–1032 (2018)
20. Wang, H., Xu, Z., Jia, S., Xia, Y., Zhang, X.: Why current differential privacy schemes are inapplicable for correlated data publishing? World Wide Web **24**(1), 1573–1413 (2021)
21. Li, Q., Gundersen, J.S., Heusdens, R., Christensen, M.G.: Privacy-preserving distributed processing: metrics, bounds and algorithms. IEEE Trans. Inf. Foren. Secur. **16**, 2090–2103 (2021)
22. Wang, C.X., Song, Y., Tay, W.P.: Arbitrarily strong utility-privacy tradeoff in multi-agent systems. IEEE Trans. Inf. Foren. Secur. **16**, 671–684 (2021)
23. Zhang, W., Li, M., Tandon, R., Li, H.: Online location trace privacy: an information theoretic approach. IEEE Trans. Inf. Foren. Secur. **14**(1), 235–250 (2019)
24. Theodorakopoulos, G., Shokri, R., Troncoso, C., Hubaux, J., Boudec, J.L.: Prolonging the hide-and-seek game: optimal trajectory privacy for location-based services. In: Proceedings of Workshop on Privacy in the Electronic Society, pp. 73–82 (2014)
25. Shokri, R., Theodorakopoulos, G., Troncoso, C.: Privacy games along location traces: a game-theoretic framework for optimizing location privacy. ACM Trans. Privacy Secur. **19**(4), 1–31 (2016)
26. Oya, S., Troncoso, C., Pérez-González, F.: Rethinking location privacy for unknown mobility behaviors. In: Proceedings of IEEE European Symposium on Security and Privacy, pp. 416–431 (2019)
27. Duchi, J.C., Jordan, M.I., Wainwright, M.J.: Local privacy and statistical minimax rates. In: Proceedings of IEEE Annual Symposium on Foundations of Computer Science, pp. 429–438 (2013)
28. Cabañas, J.G., Cuevas, Á., Cuevas, R.: Unveiling and quantifying facebook exploitation of sensitive personal data for advertising purposes. In: Proceedings of USENIX Security Symposium, pp. 479–495 (2018)
29. Liu, H., Li, X., Li, H., Ma, J., Ma, X.: Spatiotemporal correlation-aware dummy-based privacy protection scheme for location-based services. In: Proceedings of IEEE International Conference on Computer Communications, pp. 1–9 (2017)
30. Lee, B., Oh, J., Yu, H., Kim, J.: Protecting location privacy using location semantics. In: Proceedings of ACM SIGKDD International Conference on Knowledge Discovery and Data Mining, pp. 1289–1297 (2011)

31. Chen, L., Shang, S., Jensen, C.S., Yao, B., Kalnis, P.: Parallel semantic trajectory similarity join. In: Proceedings of IEEE International Conference on Data Engineering, pp. 997–1008 (2020)
32. Qian, X., Wu, Y., Li, M., Ren, Y., Jiang, S., Li, Z.: Last: location-appearance-semantic-temporal clustering based POI summarization. IEEE Trans. Multimedia **23**, 378–390 (2021)
33. Kreer, J.: A question of terminology. IRE Trans. Inf. Theory **3**(3), 208–208 (1957)
34. Permuter, H.H., Kim, Y.H., Weissman, T.: Interpretations of directed information in portfolio theory, data compression, and hypothesis testing. IEEE Trans. Inf. Theory **57**(6), 3248–3259 (2011)
35. Peled, O., Sabag, O., Permuter, H.H.: Feedback capacity and coding for the $(0, k)$-RLL input-constrained bec. IEEE Trans. Inf. Theory **65**(7), 4097–4114 (2019)
36. Ching, W.K., Ng, M.K., Fung, E.S.: Higher-order multivariate markov chains and their applications. Linear Algebra Appl. **428**(2–3), 492–507 (2008)
37. Wang, W., Ying, L., Zhang, J.: On the relation between identifiability, differential privacy, and mutual-information privacy. IEEE Trans. Inf. Theory **62**(9), 5018–5029 (2016)
38. Lei, Y., Ling, L., Pu, C.: Dynamic differential location privacy with personalized error bounds. In: Proceedings of Network and Distributed System Security Symposium, pp. 1–15 (2017)
39. Erdemir, E., Dragotti, P.L., Gündüz, D.: Privacy-aware time-series data sharing with deep reinforcement learning. IEEE Trans. Inf. Foren. Secur. **16**, 389–401 (2021)
40. Oya, S., Troncoso, C., Pérez-González, F.: Back to the drawing board: revisiting the design of optimal location privacy-preserving mechanisms. In: Proceedings of ACM SIGSAC Conference on Computer and Communications Security, pp. 1959–1972 (2017)
41. Wang, Z., Liu, W., Pang, X., Ren, J., Liu, Z., Chen, Y.: Towards pattern-aware privacy-preserving real-time data collection. In: Proceedings of IEEE International Conference on Computer Communications, pp. 109–118 (2020)
42. Hamm, J.: Enhancing utility and privacy with noisy minimax filters. In: Proceedings of IEEE International Conference on Acoustics, Speech and Signal Processing, pp. 6389–6393 (2017)
43. Han, D., Yuan, X.: A note on the alternating direction method of multipliers. J. Optim. Theory Appl. **155**(1), 227–238 (2012)
44. Karaboga, D., Akay, B.: A modified artificial bee colony (ABC) algorithm for constrained optimization problems. Appl. Soft Comput. **11**(3), 3021–3031 (2011)
45. Deb, K.: An efficient constraint handling method for genetic algorithms. Comput. Methods Appl. Mech. Eng. **186**(2), 311–338 (2000)
46. Xiao, Z., Li, F., Wu, R., Jiang, H., Hu, Y., Ren, J., Cai, C., Iyengar, A.: Trajdata: on vehicle trajectory collection with commodity plug-and-play OBU devices. IEEE Int. Things J. **7**(9), 9066–9079 (2020)
47. Yang, S., Wang, J., Fan, W., Zhang, X., Wonka, P., Ye, J.: An efficient admm algorithm for multidimensional anisotropic total variation regularization problems. In: Proceedings of ACM SIGKDD International Conference on Knowledge Discovery and Data Mining, pp. 641–649 (2013)

Chapter 4
Pricing-Aware Location Privacy Preservation in Crowdsensing

4.1 Introduction

In the last decade, ride-on-demand (RoD) services, exemplified by platforms including Uber, Lyft, and Didi, have experienced remarkable growth as key crowdsensing applications. In a typical RoD system, both drivers and riders benefit from location-based services like dynamic pricing and ride-matching by sharing their real-time locations with the RoD server. Upon receiving this location information, the server returns a ride-matching outcome and a trip fare, which is determined by applying a dynamic pricing multiplier to a fixed base rate computed from travel time and distance. The dynamic pricing mechanism is essential, as it continuously adjusts the balance between available drivers (supply) and incoming ride requests (demand) [1–7]. For instance, a higher multiplier can attract more drivers while discouraging non-urgent ride requests. Despite the convenience, RoD services raise serious privacy issues since the server can directly observe users' real locations [8–13].

To address these privacy issues, numerous state-of-the-art techniques [10, 11, 14–20] have been developed that utilize data perturbation to protect users' location privacy. These methods generally fall into two categories, i.e., local (or user-centric) [10, 14–16] and centralized [11, 17, 18] privacy-preserving techniques. Centralized approaches rely on a trusted third party to add noise perturbation, whereas local methods empower users to modify their location data. In practice, relying on a trusted third party is problematic, as it may compromise users' privacy for commercial gain and remain susceptible to malicious attacks. High-profile incidents, such as Facebook's data breach and Didi's over-collection of personal information, have further eroded public confidence in third-party-based solutions [21, 22]. Therefore, establishing a robust local privacy-preserving framework for RoD services is crucial.

Nevertheless, perturbing location data can disrupt the balance between driver supply and rider demand, reducing the quality of dynamic pricing services. For example, if the riders' locations in central business districts (CBD) are obfuscated

Z. Zheng et al., *Perturbation Based Privacy in Crowdsensing*, Wireless Networks,
https://doi.org/10.1007/978-3-031-95052-0_4

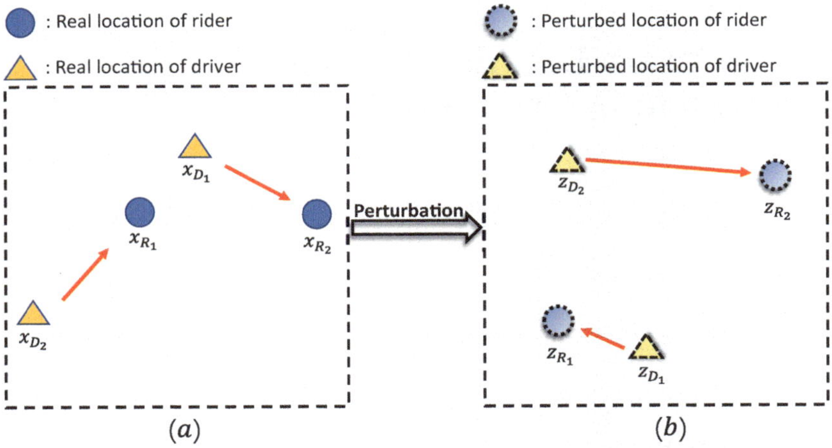

Fig. 4.1 Motivation example: the ride-matching service is susceptible to the impact of perturbations. (**a**) The optimal ride-matching result without perturbation is (D_1, R_2) and (D_2, R_1), where $d(x_{D_1}, x_{R_2}) > d(x_{D_1}, x_{R_1})$ and $d(x_{D_2}, x_{R_2}) > d(x_{D_2}, x_{R_1})$. (**b**) The optimal ride-matching result after perturbation is (D_1, R_1) and (D_2, R_2), where $d(z_{D_1}, z_{R_2}) > d(z_{D_1}, z_{R_1})$ and $d(z_{D_2}, z_{R_2}) > d(z_{D_2}, z_{R_1})$

to residential areas during evening rush hours, this could lead to inflated fares in residential regions, even though the actual demand for drivers remains higher in the central zones. Moreover, the ride-matching service is particularly sensitive to noise perturbation because it relies on accurately measuring the geo-distance between drivers and riders. As demonstrated in Fig. 4.1, even minor inaccuracies can significantly impair the quality of ride-matching service.

Distinct from existing works, we propose a perturbation-based privacy-preserving framework for RoD services that defends against an honest-but-curious server without relying on any third party, with the goal of delivering high-quality dynamic pricing and ride-matching services while protecting users' location data privacy. In pursuing this objective, we encounter three major challenges. *The first challenge is measuring the quality loss of dynamic pricing and ride-matching services caused by local perturbation.* Since perturbation is executed on the client side, the complete perturbed location dataset cannot be directly observed, making an accurate evaluation of quality loss difficult. *The second challenge is ensuring high-quality dynamic pricing and ride-matching services.* These two services are interdependent, i.e., improvements in one may negatively affect the other. This interplay makes it challenging to design a perturbation-based privacy-preserving framework that maintains the quality of both services. *The third challenge is striking an optimal balance between service quality and location privacy.* Typically, stronger perturbation enhances privacy protection but also leads to greater utility loss, making it tough to offer excellent services while effectively protecting location privacy.

To tackle these challenges, we propose a Pricing-Aware Differentially Private framework for RoD services, referred to as PADP-RoD. In our approach, we first design an adaptive grid that takes into account supply and demand variations to partition the spatiotemporal region. Based on this grid, we propose separate utility metrics for dynamic pricing and ride matching to quantify the quality degradation caused by local perturbation. Once the quality loss is assessed, we naturally resolve the conflict between ensuring high-quality services and protecting location privacy by formulating a multi-objective optimization problem, minimizing the quality losses of dynamic pricing and ride-matching services under the given privacy constraint. To achieve the optimal trade-off between service quality and privacy preservation, we decompose the formulated multi-objective problem into two sub-problems: one for dynamic pricing utility and the other for ride-matching utility. These sub-problems are solved using the dynamic pricing mapping algorithm and the ride-matching mapping algorithm, respectively. We evaluate PADP-RoD by comparing it with several state-of-the-art methods using a real-world event-log dataset. The experimental results demonstrate that PADP-RoD effectively delivers high-quality dynamic pricing and ride-matching services while resisting Bayesian inference attacks. The key contributions of this chapter are summarized as follows:

- We propose PADP-RoD, a pricing-aware differentially private framework for ride-on-demand services, enabling the RoD server to deliver high-quality dynamic pricing and ride-matching services without accessing users' real location data.
- We propose an adaptive grid that responds to supply and demand changes to capture the fluctuation of price multiplier. Based on this grid, we propose two utility metrics to quantify the service quality degradation of dynamic pricing and ride-matching arising from local perturbation.
- Utilizing our newly proposed metrics, we formulate PADP-RoD as a multi-objective minimization problem. Given its multi-objective nature, we decompose it into two sub-problems, one addressing dynamic pricing utility and the other focusing on ride-matching utility, and solve them using the dynamic pricing mapping algorithm and the ride-matching mapping algorithm, respectively. Moreover, we prove that PADP-RoD satisfies both ε_d-DP and ε_d-identifiability.

The remainder of this chapter is organized as follows. In Sect. 4.2, we introduce the system model, adversary model, and design objectives, followed by a discussion of two privacy concepts, such as differential privacy and identifiability. Sections 4.3 and 4.4 provide a detailed presentation of PADP-RoD. Specifically, Sect. 4.3 proposes two utility metrics to evaluate the quality loss induced by noise perturbation, and Sect. 4.4 formulates PADP-RoD as a multi-objective optimization problem and presents an efficient algorithm for its resolution. The theoretical analysis of PADP-RoD is provided in Sect. 4.5, and its performance evaluation is discussed in Sect. 4.6. Finally, Sect. 4.7 concludes this chapter.

4.2 Problem Statement and Basic Concepts

4.2.1 Problem Statement

We state the problem addressed in this chapter from the following three aspects: the system model, the adversary model, and the design goals.

System Model Figure 4.2 illustrates our framework for pricing-aware, privacy-preserving RoD services, which include privacy-conscious drivers and riders (a.k.a. users), and a semi-honest RoD server. In this setup, the server reliably provides high-quality services yet remains curious about users' real locations. To mitigate this privacy risk, users locally apply the PADP-RoD privacy mechanism to generate and share perturbed (or false) locations instead of their real ones. Upon receiving these modified locations, the server computes the trip fare and returns the ride-matching result (a.k.a. ride-matching service). The fare is calculated by multiplying a dynamic price multiplier, which reflects current supply and demand (a.k.a. dynamic pricing service), with a fixed base price determined by trip duration and distance. Based on the provided fare and ride-matching outcome, users then decide whether to accept the service. The key parameters and notations are summarized in Table 4.1, and location subscripts are omitted when distinguishing between drivers' and riders' locations is unnecessary.

Adversary Model After capturing the perturbed locations produced by PADP-RoD, the adversary initiates a Bayesian inference attack [23, 24] to deduce the real locations. Below, we provide a detailed definition of the Bayesian inference attack.

Definition 4.1 (Bayesian Inference Attack) We assume that the adversary's background knowledge is a prior probability distribution $\mathcal{A}_{Z|X}(z \mid x)$ and a victim's

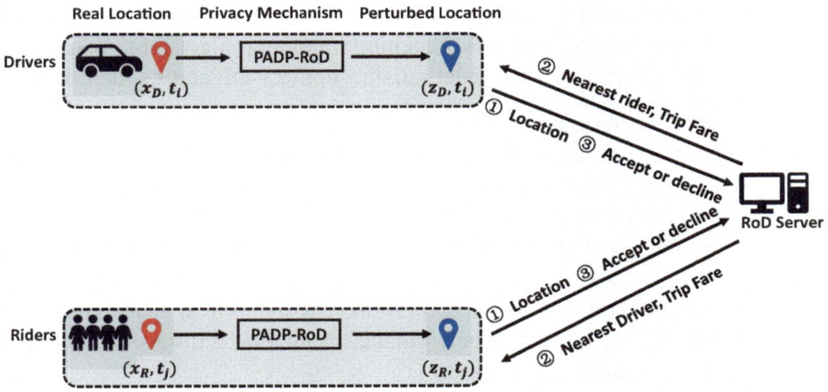

Fig. 4.2 Workflow of pricing-aware privacy-preserving RoD services

Table 4.1 Key notations and parameters

Symbol	Meaning
x_D, y_D, z_D	Real, temporary, or perturbed location of driver
x_R, y_R, z_R	Real, temporary, or perturbed location of rider
$S(*)$, $D(*)$	Supply or demand function
$\rho_s(*)$	Price multiplier elasticity in spatial dimension
$\rho_t(*)$	Price multiplier elasticity in temporal dimension
$Pri(*)$	Dynamic pricing utility metric
$Dis(*)$	Ride-matching utility metric
$\mathcal{B}_{Y\|X}$	Dynamic pricing mapping mechanism
$C_{Z\|Y}$	Ride-matching mapping mechanism

profile $p_X(\cdot)$. Given the perturbed location z, Bayes' theorem is used to compute the posterior probability of each possible real location as follows,

$$p(x \mid z) = \frac{\mathcal{A}_{Z|X}(z \mid x) \cdot p_X(x)}{\sum_{x^* \in \mathcal{L}} \mathcal{A}_{Z|X}(z \mid x^*) \cdot p_X(x^*)}, \quad \forall x \in \mathcal{L}. \tag{4.1}$$

Once these posterior probabilities are obtained, the real location corresponding to z is inferred by

$$\widehat{x} = \min_{x \in \mathcal{L}} p(x \mid z). \tag{4.2}$$

Design Goals Our objective is to enable the RoD server to offer high-quality dynamic pricing and ride-matching services while remaining unaware of users' real locations. To balance these conflicting aims, we formulate PADP-RoD as a multi-objective optimization problem,

Maximize: The service quality of dynamic pricing and ride-matching,

Subject To: The mapping mechanism $\mathcal{A}_{Z|X}$ satisfies ε-DP.

To determine the optimal mapping mechanism $\mathcal{A}_{Z|X}$, we must quantify the quality losses in dynamic pricing and ride-matching services caused by noise perturbation, and develop a specialized algorithm to solve the optimization problem. These details are provided in Sects. 4.3 and 4.4.

4.2.2 Basic Concepts

In this part, we introduce two privacy notions, i.e., differential privacy [25] and identifiability [26]. Prior to discussing these notions, we first define the concept of neighboring locations.

Definition 4.2 (Neighboring Locations) Let the RoD service area consist of a location set $\mathcal{L}$, structured in M rows and N columns. Each location $\forall x \in \mathcal{L}$ is represented as (m, n), where $m \in \{1, \cdots, M\}$ and $n \in \{1, \cdots, N\}$. Two locations $x, x' \in \mathcal{L}$ are considered neighbors if they differ in only one coordinate. For instance, locations $x := (m, n)$ and $x' := (m, n' \neq n)$ (or $x' := (m' \neq m, n)$) are neighbors. However, if both coordinates differ, i.e., $x := (m, n)$ and $x' := (m' \neq m, n' \neq n)$, they are not considered neighbors.

Differential Privacy Differential privacy (DP) is extensively employed to protect users' location data privacy by adding noise to their real locations before sharing with the server [15, 16, 27]. A randomized mechanism is considered to satisfy DP if any neighboring locations are perturbed to a given output with similar probabilities, as formally defined below.

Definition 4.3 (Differential Privacy) Let the service area of RoD include a set of locations $\mathcal{L}$. The mapping mechanism $\mathcal{A}_{Z|X}$ is said to be ε_d-DP for some $\varepsilon_d \in \mathbb{R}^+$, if for any pair of neighboring locations $\forall x, x' \in \mathcal{L}$,

$$\mathcal{A}_{Z|X}(z \mid x) \leq e^{\varepsilon_d} \cdot \mathcal{A}_{Z|X}(z \mid x'), \ \forall z \in \mathcal{L}, \tag{4.3}$$

where $\mathcal{A}_{Z|X}(z \mid x)$ is the probability of perturbing x to z.

Identifiability Identifiability, as defined in [26], is based on the Bayesian notion of indistinguishability between any two neighboring locations. Specifically, after observing a perturbed location, the posterior probabilities for any neighboring locations remain similar, making it difficult to pinpoint a user's real location. Consequently, identifiability is widely employed to defend against Bayesian inference attacks [23, 24], and its detailed definition is provided below.

Definition 4.4 (Identifiability) Let the service area of RoD includes a set of locations $\mathcal{L}$. The mapping mechanism $\mathcal{A}_{Z|X}$ satisfies ε_i-identifiability for some $\varepsilon_i \in \mathbb{R}^+$, if for any pair of neighboring locations $x, x' \in \mathcal{L}$:

$$\mathcal{A}_{X|Z}(x \mid z) \leq e^{\varepsilon_i} \cdot \mathcal{A}_{X|Z}(x' \mid z), \ \forall z \in \mathcal{L}, \tag{4.4}$$

where $\mathcal{A}_{X|Z}(x \mid z)$ is the posterior probability after observing the location z.

4.3 Utility Loss Metrics

To quantify the quality loss of dynamic pricing and ride-matching services caused by noise perturbation, we introduce the adaptive supply and demand aware grid, and then propose the dynamic pricing utility metric and the ride-matching utility metric.

4.3.1 Adaptive Supply and Demand Aware Grid

The price multiplier tends to exhibit abrupt variations in hotspot areas, such as CBD (spatial dimension) and morning peak (temporal dimension) [1, 2]. To capture these fluctuations, we propose an adaptive grid that accounts for supply and demand dynamics. Specifically, we employ finer (smaller) grid cells in hotspot regions while using coarser (larger) cells in less active areas. This approach ensures that price multipliers remain nearly uniform within each cell while preserving distinctions between adjacent cells. To this end, we define a metric called price multiplier elasticity, which measures the rate-of-changes of price multiplier in spatial and temporal dimensions. Based on this elasticity, we then propose an adaptive supply and demand aware discretization algorithm to segment the spatiotemporal region.

Since the price multiplier is inherently tied to the supply-demand dynamic, its value shifts as these dynamics change. Therefore, we define price multiplier elasticity as the absolute slope of the supply-demand relationship, with a more detailed definition provided below.

Definition 4.5 (Price Multiplier Elasticity in Spatial Dimension) Let $S(x, t)$ and $D(x, t)$ represent the supply and demand functions, respectively. The relationship between supply and demand can be formulated as $S(x, t) - D(x, t)$. In the spatial dimension, we define the price multiplier elasticity as

$$\rho_s(x, t) = \left| \frac{\partial S(x, t)}{\partial x} - \frac{\partial D(x, t)}{\partial x} \right|. \tag{4.5}$$

Definition 4.6 (Price Multiplier Elasticity in Temporal Dimension) Let $S(x, t)$ represent the supply function and $D(x, t)$ represent the demand function. Their interaction is captured by the difference $S(x, t) - D(x, t)$. In the temporal dimension, we define the price multiplier elasticity as

$$\rho_t(t) = \left| \frac{\mathrm{d}}{\mathrm{d}t} \int_{\Omega} S(x, t) - D(x, t) dx \right|. \tag{4.6}$$

When a location x is a hotspot at time t, the price multiplier elasticity $\rho_s(x, t)$ defined in Definition 4.5 increases. Similarly, if time t represents a hotspot, the elasticity $\rho_t(t)$ defined in Definition 4.6 becomes larger. Therefore, when $\rho_s(x, t)$ is high, finer grid cells should be deployed around location x at time t, and the spatial grid should be reconfigured during periods when $\rho_t(t)$ is elevated. Next, we explain how to discretize the spatiotemporal region based on these elasticity measures.

Algorithm 3 Lloyd algorithm for CVT

Input: An spatial region Ω, a positive integer g_n, a time t, and a threshold for convergence η

Output: Centroidal Voronoi tessellation $\{l_i, V_i\}_{i=1}^{g_n}$

1: Initialize a set of g_n generators $\{l_i\}_{i=1}^{g_n} \subset \Omega$;

while True **do**
  Construct the Voronoi tessellation $\{V_i\}_{i=1}^{g_n}$ of Ω according to the generators $\{l_i\}_{i=1}^{g_n}$;
  Determine the set of centroids, i.e., $\{l_i^*\}_{i=1}^{g_n}$, of Voronoi tessellation $\{V_i\}_{i=1}^{g_n}$ according to the Eq. (4.8);
  if $\sum_{i=1}^{g_n} d(l_i^*, l_i) \leq \eta$ **then**
    Update the set of generators: $\{l_i\}_{i=1}^{g_n} \leftarrow \{l_i^*\}_{i=1}^{g_n}$;
    Break;
  end if
  Update the set of generators: $\{l_i\}_{i=1}^{g_n} \leftarrow \{l_i^*\}_{i=1}^{g_n}$;
end while

Algorithm 4 Adaptive supply and demand aware discretization algorithm

Input: A spatial region Ω, a positive integer g_n, a refinement number ref, and a percentage of cells that need to be refined ref_{per}
Output: Adaptive supply and demand aware grid $\mathbb{G}$
Initialize uniform point set in the temporal region [0, 24 $\times$ 60 minutes], i.e., set T;
for $i \in [1, ref]$ **do**
  Calculate the price multiplier elasticity per temporal cell by $\int_{t_i}^{t_i+1} \rho_t(t)dt$, $\forall t_i \in T$;
  Mark top $\left|ref_{per} \cdot (|T|+1)\right|$ temporal cells of price multiplier elasticity;
  Update the temporal grid T by bisecting each marked cell;
end for
for $t_i \in T$ **do**
  $\{l_j, V_j\}_{j=1}^{g_n} \leftarrow$ Discretize Ω by Algorithm 3;
  $\mathbb{D}_i \leftarrow$ Dual tessellation of $\{l_i, V_i\}_{i=1}^{g_n}$;
  Append $\mathbb{D}_i$ to $\mathbb{G}$;
end for

To partition the spatial domain, we employ the CVDT structure [28]. Unlike traditional grid methods such as the uniform grid [29] and the two-step grid [17], the CVDT grid dynamically adjusts its cell sizes and overall layout based on the price multiplier elasticity $\rho_s(x, t)$. Below, we provide the formal definition of CVDT and its standard construction algorithm.

Definition 4.7 (Centroidal Voronoi Delaunay Triangulation) Given a spatial service area Ω and a set of generators $\{l_i\}_{i=1}^{g_n} \subset \Omega$. We define the Voronoi region V_i corresponding to generator l_i as

$$V_i = \{l \in \Omega \mid d(l, l_i) < d(l, l_j),\ \forall j \in \{1, \ldots, g_n\}/\{i\}\}, \tag{4.7}$$

where $d(l, *)$ denotes the geo-distance between locations l and $*$. It could be seen from Eq. (4.7), the Voronoi region $V_i \cap V_j = \emptyset$, $\forall j \neq i$ and $\cup_{i=1}^{g_n} \overline{V}_i = \Omega$, i.e.,

$\{V_i\}_{i=1}^{g_n}$ is a tessellation of Ω. We refer to a Voronoi tessellation $\{(x_i, V_i)\}_{i=1}^{g_n}$ of Ω as a Centroidal Voronoi Tessellation (CVT) if and only if for each generator l_i is the centroid of its corresponding region V_i, i.e.,

$$l_i = \frac{\int_{V_i} l \cdot \rho_s(l, t)dl}{\int_{V_i} \rho_s(l, t)dl}, \quad \forall i \in \{1, \cdots, g_n\}. \tag{4.8}$$

The corresponding dual tessellation is the so-called CVDT.

As described in Definition 4.7, the generators are attracted to regions with high price multiplier elasticity $\rho_s(x, t)$ (i.e., spatial hotspots), which results in a more refined cell partitioning around these areas. The Lloyd algorithm, detailed in Algorithm 3, is employed to construct the CVDT. We begin with a set of initial generators $\{l_i\}_{i=1}^{g_n}$ (line 1) that serve as the starting points. In each iteration, a Voronoi tessellation $\{V_i\}_{i=1}^{g_n}$ is generated based on the current generators $\{l_i\}_{i=1}^{g_n}$ (line 3), after which the centroid of each cell is calculated following Eq. (4.8) (line 4). These centroids then replace the previous generators (line 6). When the centroids of the cells cease to move (line 5), the process converges to yield the CVT for the spatial domain (i.e., $\{l_i, V_i\}_{i=1}^{g_n}$), and the corresponding dual tessellation forms the CVDT.

For the temporal dimension, we apply a dichotomy strategy [11] to perform discretization. In each refinement step, cells with higher price multiplier elasticity $\rho_t(t)$ are identified and then split into two equal parts. Building on this idea, we propose an adaptive supply and demand aware discretization algorithm that integrates both the CVDT and the dichotomy strategy, as outlined in Algorithm 4. Initially, a uniform grid is set up over the temporal range [0, 24 × 60 minutes] (line 1). In every iteration along the temporal dimension (lines 2–6), the elasticity for each temporal cell, defined as the interval (t_i, t_{i+1}) for each $t_i \in T$, is computed using $\int_{t_i}^{t_{i+1}} \rho_t(t)dt$ (line 3). The cells with the highest elasticity values, constituting the top $ref_{per} \cdot (|T| + 1)$, are marked for refinement (line 4) and then bisected (line 5). After ref iterations, this process yields an adaptive temporal grid $T = \{t_1, t_2, \cdots\}$. Finally, the spatial region is re-discretized at each time $\forall t_i \in T$ using Algorithm 3 (lines 7–11), resulting in an adaptive supply and demand aware grid $\mathbb{G} = \{\mathbb{D}_i \mid \forall t_i \in T\}$.

4.3.2 Dynamic Pricing Utility Metric

In local scenarios, it is difficult to evaluate utility loss at the dataset level because perturbations are applied locally and the whole location dataset remains unobserved. To address this limitation, we propose a dynamic pricing utility metric based on grid $\mathbb{G}$ that measures utility loss from a user-level perspective. In essence, maintaining the supply-demand balance within each grid cell preserves the dynamic pricing utility. As shown in Fig. 4.3, if a perturbation keeps both the real location and its

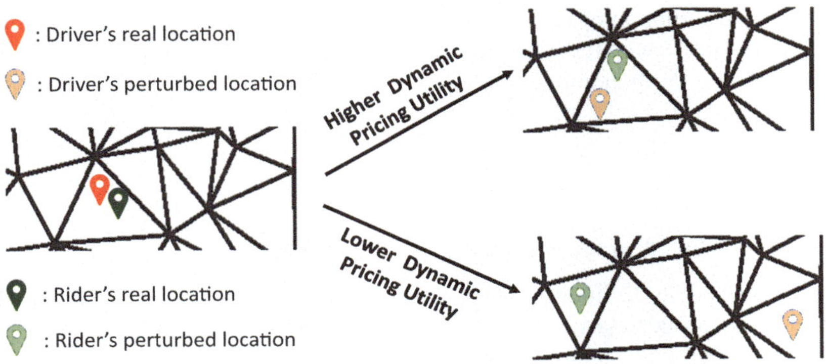

Fig. 4.3 An intuitive example of measuring the dynamic pricing utility. Perturbation occurring within the same cell results in higher dynamic pricing utility, whereas perturbation that moves real locations to other cells leads to lower dynamic pricing utility. Additionally, when the dynamic pricing utility is preserved, perturbation still leads to inaccurate geo-distance between the driver's location and the rider's location

perturbed version within the same cell, the utility remains intact; conversely, if the perturbation shifts the location into a different cell, the utility may be compromised. This method is particularly effective in local settings, as intra-cell perturbations do not disturb the cell's inherent supply-demand relationship. Next, we present a detailed definition of the dynamic pricing utility metric.

Definition 4.8 (Dynamic Pricing Utility Metric) Let $\forall x \in \mathcal{L}$ (or $\forall z \in \mathcal{L}$) is the real (or perturbed) location, $\mathcal{A}_{Z|X}(z \mid x)$ is a randomized obfuscation mechanism of perturbing x to z, and $\mathbb{G} = \{\mathbb{D}_i \mid \forall t_i \in T\}$ is the adaptive supply and demand aware grid. Given time (t_i, t_{i+1}), the dynamic pricing utility metric is defined as

$$Pri(\mathcal{A}_{Z|X}) := \sum_{x,z \in \mathcal{L}} p_X(x) \cdot \mathcal{A}_{Z|X}(z \mid x) \cdot \mathbb{1}_{\mathbb{D}_i(z) \neq \mathbb{D}_i(x)}. \tag{4.9}$$

The indicator function $\mathbb{1}_{\mathbb{D}_i(z) \neq \mathbb{D}_i(x)}$ is formulated as

$$\mathbb{1}_{\mathbb{D}_i(z) \neq \mathbb{D}_i(x)} = \begin{cases} 1 & \text{if } \mathbb{D}_i(z) \neq \mathbb{D}_i(x) \\ 0 & \text{if } \mathbb{D}_i(z) = \mathbb{D}_i(x) \end{cases}, \tag{4.10}$$

where $\mathbb{D}_i(*)$ returns the cell of location $*$ in the grid $\mathbb{D}_i$ during (t_i, t_{i+1}). For example, $\mathbb{D}_i(*) = C_j^i$ ($C_j^i \in \mathbb{D}_i$) means that the location $*$ is located in the spatial cell $C_j^i = (m, n)$ ($m \in \{1, \cdots, M\}, n \in \{1, \cdots, N\}$) during (t_i, t_{i+1}).

According to Definition 4.8, a higher dynamic pricing utility metric indicates a greater loss in utility. By minimizing this metric under the privacy constraints, we can effectively preserve the utility of dynamic pricing. In doing so, the RoD server is able to deliver high-quality dynamic pricing services without needing access to users' real locations.

4.3.3 *Ride-Matching Utility Metric*

Ride-matching is a fundamental component of the RoD service [30], yet applying noise perturbation tends to diminish its service quality. This degradation occurs because the geo-distance between perturbed locations often deviates from that between the real locations, i.e., $d(x_R, x_D) \neq d(z_R, z_D)$. As depicted in Fig. 4.3, even when perturbations remain confined within the same cell, they do not necessarily preserve comparable geo-distances. Consequently, ensuring dynamic pricing utility alone is insufficient to maintain the quality of ride-matching service.

Before defining the ride-matching utility metric, we approximate the geo-distance between real locations by calculating the expected travel distance between their perturbed versions. As illustrated in Fig. 4.4, the geo-distance $d(x_R, x_D)$ is approximated by the expected travel distance $d^*(z_R, z_D)$, which is the sum of four components: $z_R \leftarrow x_{1,R} — x_{1,D} \rightarrow z_D$, $z_R \leftarrow x_{1,R} — x_{2,D} \rightarrow z_D$, $z_R \leftarrow x_{2,R} — x_{1,D} \rightarrow z_D$, and $z_R \leftarrow x_{2,R} — x_{2,D} \rightarrow z_D$. To clarify, consider the first component $z_R \leftarrow x_{1,R} — x_{1,D} \rightarrow z_D$. Based on the mapping mechanism $\mathcal{A}_{Z|X}$ and the probability $p_X(x_{1,R})$ (which is inferred from historical location data), the perturbed location z_D equals the real location $x_{1,R}$ with probability $p_X(x_{1,R}) \times \mathcal{A}(z_R \mid x_{1,R})$. Similarly, z_D corresponds to $x_{1,D}$ with probability $p_X(x_{1,D}) \times \mathcal{A}(z_D \mid x_{1,D})$. Thus, we have $d(z_R, z_D) = d(x_{1,R}, x_{1,D})$ with probability $p_X(x_{1,R}) \times p_X(x_{1,D}) \times \mathcal{A}(z_R \mid x_{1,R}) \times \mathcal{A}(z_D \mid x_{1,D})$. This allows us to define the component as $z_R \leftarrow x_{1,R} — x_{1,D} \rightarrow z_D := p_X(x_{1,R}) \times p_X(x_{1,D}) \times \mathcal{A}(z_R \mid x_{1,R}) \times \mathcal{A}(z_D \mid x_{1,D}) \times d(x_{1,R}, x_{1,D})$. By applying similar reasoning to the remaining components, we arrive at $d^*(z_R, z_D) = \sum_{x_D, x_R \in \mathcal{L}} p_X(x_R) \times p_X(x_D) \times \mathcal{A}(z_R \mid x_R) \times \mathcal{A}(z_D \mid x_D) \times d(x_R, x_D)$. A detailed definition of the expected travel distance is provided below.

Definition 4.9 (Expected Travel Distance) Let $\forall z_D \in \mathcal{L}$ (or $\forall z_R \in \mathcal{L}$) is the driver's (or rider's) perturbed location, and $x_D \in \mathcal{L}$ (or $x_R \in \mathcal{L}$) is the real version of

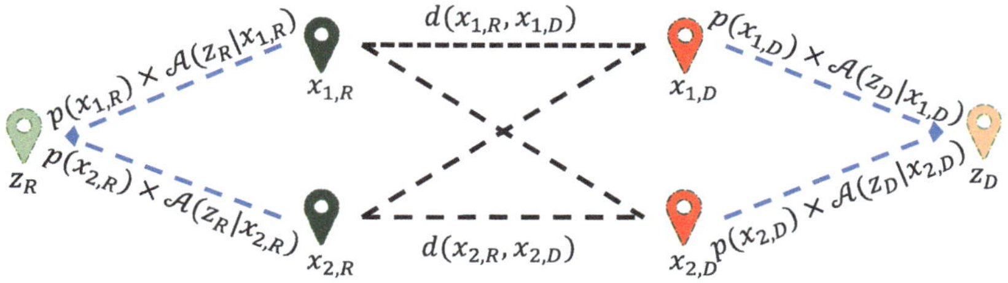

Fig. 4.4 An intuitive example of calculating the expected travel distance, where $\mathcal{L}$ is formed by two rectangles of equal size (i.e., $|\mathcal{L}| = 2$). The expected travel distance between perturbed locations z_R and z_D (i.e., $d^*(z_R, z_D)$) is the sum of four parts: $z_R \leftarrow x_{1,R} — x_{1,D} \rightarrow z_D$, $z_R \leftarrow x_{1,R} — x_{2,D} \rightarrow z_D$, $z_R \leftarrow x_{2,R} — x_{1,D} \rightarrow z_D$, and $z_R \leftarrow x_{2,R} — x_{2,D} \rightarrow z_D$, where $z_R \leftarrow x_{1,R} — x_{1,D} \rightarrow z_D$ denotes $d(z_R, z_D) = d(x_{1,R}, x_{1,D})$ with probability $p_X(x_{1,R}) \times \mathcal{A}(z_R \mid x_{1,R}) \times p_X(x_{1,D}) \times \mathcal{A}(z_D \mid x_{1,D})$

perturbed location z_D (or z_R). Given the mapping mechanism $\mathcal{A}_{Z|X}$ and the profile $p_X(\cdot)$, the expected travel distance between z_D and z_R is defined as

$$\begin{aligned} & d^*(z_D, z_R) \\ & = \sum_{x_D, x_R \in \mathcal{L}} p_X(x_D) \cdot p_X(x_R) \cdot \mathcal{A}_{Z|X}(z_D \mid x_D) \cdot \mathcal{A}_{Z|X}(z_R \mid x_R) \cdot d(x_D, x_R), \end{aligned} \tag{4.11}$$

where $d(x_D, x_R)$ is the geo-distance between real locations x_D and x_R.

The RoD server can directly compute the expected travel distance $d^*(\cdot)$, which is used in place of the real geo-distance $d(\cdot)$ in the ride-matching service. However, using $d^*(z_D, z_R)$ to approximate the real geo-distance $d(x_D, x_R)$ introduces a significant error, as indicated by $|d(x_D, x_R) - d^*(z_D, z_R)| \gg 0$. By measuring this error, thus, we quantify the ride-matching utility loss due to noise perturbation. The detailed definition is provided below.

Definition 4.10 (Ride-Matching Utility Metric (RMUM)) Let $\forall x_D \in \mathcal{L}$ (its perturbed version is $z_D \in \mathcal{L}$) is the driver's real location and $\forall x_R \in \mathcal{L}$ (its perturbed version is $z_R \in \mathcal{L}$) is the rider's real location. Then, the ride-matching utility metric is defined as

$$Dis(\mathcal{A}_{Z|X}) := \sum_{x_D, x_R \in \mathcal{L}} \left| d(x_D, x_R) - d^*(z_D, z_R) \right|, \tag{4.12}$$

where $d(\cdot)$ is the geo-distance and the expected travel distance $d^*(\cdot)$ is defined in Definition 4.9.

Definition 4.10 indicates that a higher value of the ride-matching utility metric corresponds to a greater utility loss. Accordingly, minimizing this metric under the privacy constraint helps preserve ride-matching quality. In essence, even though the RoD server only receives perturbed location data, it still manages to deliver high-quality ride-matching service.

4.4 Pricing-Aware Privacy Mapping Mechanism

Leveraging the newly defined utility metrics for dynamic pricing (see Definition 4.8) and ride-matching (see Definition 4.10), we reformulate PADP-RoD into a multi-objective optimization framework, denoted as (O), which minimizes utility loss while adhering to differential privacy constraints (as described in Sect. 4.4.1). This approach enables the RoD server to deliver superior dynamic pricing and ride-matching services without leaking users' location privacy. To achieve this, we solve (O) in Sect. 4.4.2 by identifying the optimal privacy mapping mechanism. In particular, as illustrated in Fig. 4.5, we obtain high-quality dynamic pricing service by confining perturbations to within the same cell (using the mapping mechanism

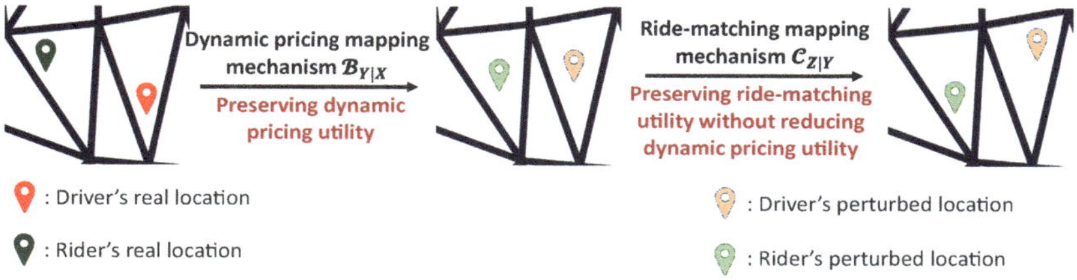

Fig. 4.5 The workflow of PADP-RoD. The dynamic pricing mapping mechanism $\mathcal{B}_{Y|X}$ restricts perturbations from occurring within the same cell (ensuring that a driver's real and perturbed locations are in the same cell), while the ride-matching mapping mechanism $C_{Z|Y}$ only fine-tune users' locations within that same cell. By this perturbation mechanism, we can obtain high-quality dynamic pricing and ride-matching services

$\mathcal{B}_{Y|X}$). Subsequently, by fine-tuning these perturbed locations within the cell, we maintain the service quality for both dynamic pricing and ride-matching (using the mapping mechanism $C_{Z|Y}$). This strategy effectively decomposes the original multi-objective problem over the mapping mechanism $\mathcal{A}_{Z|X}$ into two single-objective problems, i.e., problem (O_d) that focuses on $\mathcal{B}_{Y|X}$, and problem (O_r) that centers on $C_{Z|Y}$. Finally, as detailed in Sect. 4.4.2, we resolve (O_d) and (O_r) using our proposed dynamic pricing and ride-matching mapping algorithms, respectively.

4.4.1 Constructing Optimization Model

Intuitively, there is an inverse relationship between the desired level of privacy and the utility received, i.e., higher privacy leads to lower utility, and vice versa. Consequently, we formulate the muti-objective optimization problem given in Sect. 4.2.1 as

$$(O): \min_{\mathcal{A}_{Z|X}(z|x)} \left(Pri(\mathcal{A}_{Z|X}), Dis(\mathcal{A}_{Z|X})\right), \tag{4.13a}$$

$$\text{s.t. } \mathcal{A}_{Z|X}(z \mid x) \leq e^{\varepsilon_d} \cdot \mathcal{A}_{Z|X}(z \mid x'), \ \forall x, x' : x \sim x', z \in \mathcal{L}, \tag{4.13b}$$

$$\sum_{z \in \mathcal{L}} \mathcal{A}_{Z|X}(z \mid x) = 1, \ \forall x \in \mathcal{L}, \tag{4.13c}$$

$$\mathcal{A}_{Z|X}(z \mid x) \geq 0, \ \forall x, z \in \mathcal{L}, \tag{4.13d}$$

where ε_d denotes the privacy budget, and the mapping mechanism $\mathcal{A}_{Z|X}$ is represented as a $|\mathcal{L}| \times |\mathcal{L}|$-dimension matrix. By solving the problem (O), we can obtain the optimal local mapping mechanism $\mathcal{A}_{Z|X}(z \mid x)$. Specifically, PADP-

RoD adopts a user-centric approach that locally converts the real location x into a perturbed location z by the mapping mechanism $\mathcal{A}_{Z|X}(z \mid x)$.

4.4.2 Solving Optimization Model

The problem ($\mathcal{O}$) is inherently multi-objective, making a direct and efficient solution impractical. To address this, we decompose ($\mathcal{O}$) into two single-objective sub-problems that can be solved separately. Specifically, we decompose the mapping mechanism $\mathcal{A}_{Z|X}(z \mid x)$ into two sequential stages by

$$\mathcal{A}_{Z|X}(z \mid x) = \sum_{y \in \mathcal{L}} C_{Z|Y}(z \mid y) \cdot \mathcal{B}_{Y|X}(y \mid x), \tag{4.14}$$

where $\mathcal{B}_{Y|X}(y \mid x)$ represents the dynamic pricing mapping mechanism, and $C_{Z|Y}(z \mid y)$ corresponds to the ride-matching mapping mechanism. As illustrated in Fig. 4.5, $\mathcal{B}_{Y|X}$ is designed to preserve dynamic pricing utility by ensuring that perturbations occur within the same cell. Then, $C_{Z|Y}$ refines these adjustments to maintain both ride-matching and dynamic pricing utilities. By this way, we relax the original problem into two sub-problems: a dynamic pricing utility sub-problem (O_d), which focuses on optimizing $\mathcal{B}_{Y|X}$, and a ride-matching utility sub-problem (O_r), which optimizes $C_{Z|Y}$ based on the solution of $\mathcal{B}_{Y|X}$.

The quality of both dynamic pricing and ride-matching services depends on the mapping mechanisms $\mathcal{B}_{Y|X}$ and $C_{Z|Y}$. By solving the sub-problem (O_d), we achieve the minimum dynamic pricing utility loss, denoted as $\arg\min Pri(\mathcal{B}_{Y|X})$. Next, to ensure that dynamic pricing service is not compromised, we impose the constraint $pri(C_{Z|Y}) \leq \arg\min Pri(\mathcal{B}_{Y|X}) + \Phi$ on sub-problem (O_r), where Φ represents the maximum allowable additional utility loss due to $C_{Z|Y}$. However, if $\arg\min Pri(\mathcal{B}_{Y|X})$ is unknown, it is not feasible to set an appropriate constraint for dynamic pricing utility loss in (O_r).

Dynamic Pricing Utility Sub-Problem We formulate the dynamic pricing utility sub-problem as

$$(O_d): \quad \min_{\mathcal{B}_{Y|X}(y|x)} \sum_{x,y \in \mathcal{L}} p_X(x) \cdot \mathcal{B}_{Y|X}(y \mid x) \cdot \mathbb{1}_{\mathbb{D}_i(y) \neq \mathbb{D}_i(x)}, \tag{4.15a}$$

$$\text{s.t. } \mathcal{B}_{Y|X}(y \mid x) \leq e^{\varepsilon_d} \cdot \mathcal{B}_{Y|X}(y \mid x'), \ \forall x, x' : x \sim x', y \in \mathcal{L}, \tag{4.15b}$$

$$\sum_{y \in \mathcal{L}} \mathcal{B}_{Y|X}(y \mid x) = 1, \ \forall x \in \mathcal{L}, \tag{4.15c}$$

$$\mathcal{B}_{Y|X}(y \mid x) \geq 0, \ \forall x, y \in \mathcal{L}, \tag{4.15d}$$

where the mapping mechanism $\mathcal{B}_{Y|X}$ is a $|\mathcal{L}| \times |\mathcal{L}|$-dimension matrix. Note that $p_X(x) \cdot \mathcal{B}_{Y|X}(y \mid x) = \mathcal{B}_{X,Y}(x, y) = p_Y(y) \cdot \mathcal{B}_{X|Y}(x \mid y)$, so the object (4.15a) has the following equivalent form, i.e.,

$$\sum_{x,y \in \mathcal{L}} p_X(x)\mathcal{B}_{Y|X}(y \mid x)\mathbb{1}_{\mathbb{D}_i(y) \neq \mathbb{D}_i(x)} = \sum_{x,y \in \mathcal{L}} p_Y(y)\mathcal{B}_{X|Y}(x \mid y)\mathbb{1}_{\mathbb{D}_i(y) \neq \mathbb{D}_i(x)}. \tag{4.16}$$

Since [31, Theorem 1] prove that DP and identifiability (defined in Definition 4.4) imply each other, i.e., if the mapping mechanism $\mathcal{B}_{Y|X}$ satisfies ε_d-DP then it also satisfies ε_d-identifiability; similarly if the mapping mechanism $\mathcal{B}_{Y|X}$ satisfies ε_d-identifiability then it also satisfies ε_d-DP. Thus, the constraint of ε_d-DP, i.e., Eq. (4.15b), could be equivalently replaced by the constraint of ε_d-identifiability, i.e.,

$$\mathcal{B}_{X|Y}(x \mid y) \leq e^{\varepsilon_d} \cdot \mathcal{B}_{X|Y}(x' \mid y), \ \forall x, x' : x \sim x', y \in \mathcal{L}. \tag{4.17}$$

In summary, we convert the sub-problem (O_d) to its equivalent form (O_d^1), i.e.,

$$(O_d^1): \min_{\mathcal{B}_{X|Y}, p_Y} \sum_{x,y \in \mathcal{L}} p_Y(y) \cdot \mathcal{B}_{X|Y}(x \mid y) \cdot \mathbb{1}_{\mathbb{D}_i(y) \neq \mathbb{D}_i(x)}, \tag{4.18a}$$

$$\text{s.t. } \mathcal{B}_{X|Y}(x \mid y) \leq e^{\varepsilon_d} \cdot \mathcal{B}_{X|Y}(x' \mid y), \ \forall x, x' : x \sim x', y \in \mathcal{L}, \tag{4.18b}$$

$$\sum_{x \in \mathcal{L}} \mathcal{B}_{X|Y}(x \mid y) = 1, \ \forall y \in \mathcal{L}, \tag{4.18c}$$

$$\mathcal{B}_{X|Y}(x \mid y) \geq 0, \ \forall x, y \in \mathcal{L}, \tag{4.18d}$$

$$\sum_{y \in \mathcal{L}} p_Y(y) = 1, \tag{4.18e}$$

$$p_Y(y) \geq 0, \ \forall y \in \mathcal{L}. \tag{4.18f}$$

According to [26, Lemma 1], for any fixed $y \in \mathcal{L}$, the optimal solution of (O_d^1) is

$$\mathcal{B}_{X|Y}(x \mid y) = \frac{e^{-\varepsilon_d \cdot \mathbb{1}_{\mathbb{D}_i(y) \neq \mathbb{D}_i(x)}}}{\left(1 + (|\mathcal{L}| - 1) \cdot e^{-\varepsilon_d}\right)^2}, \ \forall x \in \mathcal{L}. \tag{4.19}$$

Equation (4.19) shows that higher probability is assigned when $\mathbb{D}_i(x) = \mathbb{D}_i(y)$ and lower when they differ. Noting that

$$\mathcal{B}_{X|Y}(x \mid y) = \frac{\mathcal{B}_{X,Y}(x, y)}{p_Y(y)} = \frac{\mathcal{B}_{Y|X}(y \mid x) \cdot p_X(x)}{p_Y(y)}, \tag{4.20}$$

we finally obtain the optimal solution for sub-problem (O_d), i.e.,

$$\mathcal{B}_{Y|X}(y \mid x) = \frac{p_Y(y) \cdot e^{-\varepsilon_d \cdot \mathbb{1}_{\mathbb{D}_i(y) \neq \mathbb{D}_i(x)}}}{p_X(x) \cdot \left(1 + (|\mathcal{L}| - 1) \cdot e^{-\varepsilon_d}\right)^2}, \; \forall x, y \in \mathcal{L}, \tag{4.21}$$

Algorithm 5 Dynamic pricing mapping algorithm

Input: θ_1: Threshold for convergence, $p_X(x)$: Probability of real location
Output: $\mathcal{B}_{Y|X}(y \mid x)$: Dynamic pricing mapping mechanism

Initialize the dynamic pricing mapping mechanism $\mathcal{B}^0_{Y|X}(y \mid x)$ as a uniform distribution, i.e., $\mathcal{B}^0_{Y|X}(y \mid x) = \begin{pmatrix} 1/|\mathcal{L}| & \cdots & 1/|\mathcal{L}| \\ \vdots & \ddots & \vdots \\ 1/|\mathcal{L}| & \cdots & 1/|\mathcal{L}| \end{pmatrix}_{|\mathcal{L}|*|\mathcal{L}|}$;
Calculate $p_Y(y)$ using $\mathcal{B}^0_{Y|X}(y \mid x)$ by Eq. (4.23);
$k \leftarrow 1$;
while True **do**
  Calculate $\mathcal{B}^k_{Y|X}(y \mid x)$ using $p_Y(y)$ by Eq. (4.21);
  Calculate $p_Y(y)$ using $\mathcal{B}^k_{Y|X}(y \mid x)$ by Eq. (4.23);
  Normalize $\mathcal{B}^k_{Y|X}(y \mid x)$ by Eq. (4.24);
  if $\|\mathcal{B}^k_{Y|X}(y \mid x) - \mathcal{B}^{k-1}_{Y|X}(y \mid x)\|_p \leq \theta_1$ **then**
    $\mathcal{B}_{Y|X}(y \mid x) \leftarrow \mathcal{B}^k_{Y|X}(y \mid x)$;
    Break;
  end if
  $k \leftarrow k + 1$;
end while

Based on the previous discussion, we propose a dynamic pricing mapping algorithm (see Algorithm 5) to derive the mapping mechanism $\mathcal{B}_{Y|X}$. We initialize with a uniform mapping, i.e.,

$$\mathcal{B}_{Y|X}(y \mid x) = \begin{pmatrix} 1/|\mathcal{L}| & \cdots & 1/|\mathcal{L}| \\ \vdots & \ddots & \vdots \\ 1/|\mathcal{L}| & \cdots & 1/|\mathcal{L}| \end{pmatrix}_{|\mathcal{L}|*|\mathcal{L}|}. \tag{4.22}$$

Then, we iterate the following three steps in order:

- Probability Update: With the mapping mechanism $\mathcal{B}_{Y|X}$ fixed, we compute the probability $p_Y(y)$ using

$$p_Y(y) = \sum_{x \in \mathcal{L}} p_X(x) \cdot \mathcal{B}_{Y|X}(y \mid x), \; \forall y \in \mathcal{L}. \tag{4.23}$$

- Mechanism Update: With the updated $p_Y(y)$, we refresh the mapping mechanism $\mathcal{B}_{Y|X}(y \mid x)$ according to Eq. (4.21).
- Normalization: We normalize the updated mechanism $\mathcal{B}_{Y|X}(y \mid x)$ by

$$\mathcal{B}_{Y|X}(y \mid x) = \frac{\mathcal{B}_{Y|X}(y \mid x)}{\sum_{y^* \in \mathcal{L}} \mathcal{B}_{Y|X}(y^* \mid x)}, \ \forall x, y \in \mathcal{L}. \tag{4.24}$$

Through iterating these steps, the Eq. (4.15a) will be minimized and so does the dynamic pricing utility loss.

Ride-Matching Utility Sub-problem The ride-matching utility sub-problem, denoted as (O_r), is formulated as follows,

$$(O_r): \min_{C_{Z|Y}(z|y)} \sum_{y_D, y_R \in \mathcal{L}} \left| d(y_D, y_R) - d^*(z_D, z_R) \right|, \tag{4.25a}$$

$$\text{s.t. } C_{Z|Y}(z \mid y) \le e^{\varepsilon_d/2} \cdot C_{Z|Y}(z \mid y'), \ \forall y, y' : y \sim y', z \in \mathcal{L}, \tag{4.25b}$$

$$Pri(C_{Z|Y}) \le \arg\min Pri(\mathcal{B}_{Y|X}) + \Phi, \tag{4.25c}$$

$$\sum_{z \in \mathcal{L}} C_{Z|Y}(z \mid y) = 1, \ \forall y \in \mathcal{L}, \tag{4.25d}$$

$$C_{Z|Y}(z \mid y) \ge 0, \ \forall y, z \in \mathcal{L}, \tag{4.25e}$$

where the mapping mechanism $C_{Z|Y}$ is a $|\mathcal{L}| \times |\mathcal{L}|$-dimension matrix. Using the ride-utility mapping $C_{Z|Y}$ together with the profile $p_Y(y)$, the expected travel distance $d^*(z_D, z_R)$ (as defined in Definition 4.9) is reformulated as

$$d^*(z_D, z_R) = \sum_{y_D, y_R \in \mathcal{L}} p_Y(y_D) \cdot p_Y(y_R) \cdot C_{Z|Y}(z_D \mid y_D) \cdot C_{Z|Y}(z_R \mid y_R) \cdot d(y_D, y_R). \tag{4.26}$$

The constraint in (4.25c) serves to bound the additional dynamic pricing utility loss introduced by $C_{Z|Y}$. In Eq. (4.25c), $\arg\min Pri(\mathcal{B}_{Y|X})$ denotes the minimum utility loss achievable by the mechanism $\mathcal{B}_{Y|X}$ (obtained by solving sub-problem (O_d)), while Φ sets an upper bound on the extra loss attributed to $C_{Z|Y}$.

Since the sub-problem (O_r) is a semi-infinite programming problem [32] and difficult to solve directly, we convert it into an unconstrained optimization problem using the exact penalty function method [33]. With an appropriate penalty parameter, the solution of the exact penalty function will converge to the optimal solution of the original optimization problem. Below, we provide a detailed definition of the l_1-penalty function, a widely used exact penalty method, for sub-problem (O_r).

Definition 4.11 **(l_1-Penalty Function)** The l_1-penalty function of problem (O_r) is formulated as

$$\begin{aligned} & P(C_{Z|Y}(z \mid y), \delta) \\ = & \sum_{y_D, y_R \in \mathcal{L}} |d(y_D, y_R) - d^*(z_D, z_R)| \\ & + \delta \cdot (\sum_{y \in \mathcal{L}} |\mathcal{H}(y)| + \sum_{y, y': y \sim y', z \in \mathcal{L}} \mathcal{G}(y, y', z) + \sum_{y, z \in \mathcal{L}} \mathcal{I}(y, z) + \mathcal{K}(C_{Z|Y}, \Phi)), \end{aligned} \tag{4.27}$$

where the constant $\delta > 0$ is the penalty factor, and the penalty terms $\mathcal{H}(y)$, $\mathcal{G}(y, y', z)$, $\mathcal{I}(y, z)$ are formulated as

$$\mathcal{H}(y) = \sum_{z \in \mathcal{L}} C_{Z|Y}(z \mid y) - 1, \tag{4.28}$$

$$\mathcal{G}(y, y', z) = \max \left\{ C_{Z|Y}(z \mid y) - e^{\varepsilon_d} \cdot C_{Z|Y}(z \mid y'), 0 \right\}, \tag{4.29}$$

$$\mathcal{I}(y, z) = \max \left\{ -C_{Z|Y}(z \mid y), 0 \right\}, \tag{4.30}$$

$$\mathcal{K}(C_{Z|Y}, \Phi) = \max \left\{ Pri(C_{Z|Y}) - \arg\min Pri(\mathcal{B}_{Y|X}) - \Phi, 0 \right\}. \tag{4.31}$$

We can obtain the mapping mechanism $C_{Z|Y}$ by solving optimization problem $\min_{C_{Z|Y}} P(C_{Z|Y}, \delta)$ when given a suitable penalty factor δ. However, because the function $P(C_{Z|Y}, \delta)$ is non-differentiable (as a result of Eqs. (4.29)–(4.31)), gradient-based optimization algorithms cannot be applied. Consequently, we employ Particle Swarm Optimization (PSO) [34], a popular swarm intelligence algorithm, to tackle this challenge. Based on these observations, we propose a ride-matching mapping algorithm that integrates the l_1-penalty function method with PSO to solve the sub-problem (O_r).

Algorithm 6 Ride-matching mapping algorithm

Input: θ_2: Threshold for convergence, $\delta > 0$: Penalty factor, $\rho > 1$: Penalty factor growth coefficient, S: Swarm size, Φ: Upper bound of pricing utility loss
Output: $C_{Z|Y}(z \mid y)$: Ride-matching mapping mechanism

1: **for** each particle $k \in [1, S]$ **do**
2: Generate the position $C_{Z|Y}^{0,k}$ and the velocity $V^{0,k}$ randomly;
3: Normalize $C_{Z|Y}^{0,k}$ by Eq. (4.33);
4: **end for**
5: $Gbest^0 \leftarrow C_{Z|Y}^{0,1}$
6: **for** each particle $k \in [1, S]$ **do**
7: **if** $P(C_{Z|Y}^{0,k}, \delta) < P(Gbest^0, \delta)$ **then**
8: $Gbest^0 \leftarrow C_{Z|Y}^{0,k}$
9: **end if**
10: **end for**

11: $Gbest^1$, $\{C_{Z|Y}^{1,1}, \cdots, C_{Z|Y}^{1,S}\}$, $\{V^{1,1}, \cdots, V^{1,S}\} \leftarrow \min_{C_{Z|Y}} P(C_{Z|Y}, \delta)$ is solved by Algorithm 7 with $\{C_{Z|Y}^{0,1}, \cdots, C_{Z|Y}^{0,S}\}$ and $\{V^{0,1}, \cdots V^{0,S}\}$;
12: $C_{Z|Y}(z \mid y) \leftarrow Gbest^1$;
13: $j \leftarrow 1$;
14: **while** $\|Gbest^{j+1} - Gbest^j\|_p > \theta_2$ **do**
15: $\quad \delta \leftarrow \rho \cdot \delta$;
16: $\quad Gbest^{j+1}$, $\{C_{Z|Y}^{j+1,1}, \cdots, C_{Z|Y}^{j+1,S}\}$, $\{V^{j+1,1}, \cdots, V^{j+1,S}\} \leftarrow \min_{C_{Z|Y}} P(C_{Z|Y}, \delta)$ is solved by Algorithm 7 with $\{C_{Z|Y}^{j,1}, \cdots C_{Z|Y}^{j,S}\}$ and $\{V^{j,1}, \cdots V^{j,S}\}$;
17: $\quad C_{Z|Y}(z \mid y) \leftarrow Gbest^{j+1}$;
18: $\quad j \leftarrow j + 1$;
19: **end while**

Referring to Algorithm 6, we present the details of the ride-matching mapping algorithm. First, we randomly generate S particles (lines 1–4) that navigate a search space of dimensions $|\mathcal{L}| \times |\mathcal{L}|$ in search of an optimal solution. In the initial (0-th) swarm, each of the S particles is characterized by a position matrix $C_{Z|Y}^{0,k}$, $\forall k \in [1, S]$ (position of k-th particle in the 0-th swarm) and a corresponding velocity matrix $V^{0,k}$, $\forall k \in [1, S]$ (velocity of k-th particle in the 0-th swarm). Each candidate solution $C_{Z|Y}^{0,k}$ is initialized as

$$C_{Z|Y}^{0,k} = \begin{pmatrix} c_{1,1}^{0,k} & \cdots & c_{1,|\mathcal{L}|}^{0,k} \\ \vdots & \ddots & \vdots \\ c_{|\mathcal{L}|,1}^{0,k} & \cdots & c_{|\mathcal{L}|,|\mathcal{L}|}^{0,k} \end{pmatrix}_{|\mathcal{L}|\times|\mathcal{L}|,} \quad \forall k \in [1, S], \tag{4.32}$$

with each element $c_{m,n}^{0,k}$ drawn from the uniform distribution $U(0, 1)$. These elements are then normalized by

$$c_{m,n}^{0,k} \leftarrow \frac{c_{m,n}^{0,k}}{\sum_{\forall n\in[1,|\mathcal{L}|]} c_{m,n}^{0,k}}, \quad \forall m, n \in [1, |\mathcal{L}|]. \tag{4.33}$$

Likewise, the velocity matrix $V^{0,k}$ for each particle is initialized as

$$V^{0,k} = \begin{pmatrix} v_{1,1}^{0,k} & \cdots & v_{1,|\mathcal{L}|}^{0,k} \\ \vdots & \ddots & \vdots \\ v_{|\mathcal{L}|,1}^{0,k} & \cdots & v_{|\mathcal{L}|,|\mathcal{L}|}^{0,k} \end{pmatrix}_{|\mathcal{L}|\times|\mathcal{L}|,} \quad \forall k \in [1, S], \tag{4.34}$$

with each $v_{m,n}^{0,k}$ drawn from uniform distribution $U(0, 0.1)$. Subsequently, we identify the best initial position $Gbest^0$ from 0-th swarm (lines 5 to 10), which serves as the optimal solution candidate for (O_r) in that swarm. In each iteration, the penalty coefficient is updated by $\delta = \rho \cdot \delta$ (line 15). Then, using the current swarm, comprising the position matrices $\{C_{Z|Y}^{j,1}, \cdots, C_{Z|Y}^{j,S}\}$ and velocity matrices $\{V^{j,1}, \cdots, V^{j,S}\}$, we solve the problem $\min_{C_{Z|Y}} P(C_{Z|Y}, \delta)$ via PSO (line 16) as detailed in Algorithm 7. The best position $Gbest^{j+1}$ found in the $(j+1)$-th swarm is then assigned to the mapping mechanism $C_{Z|Y}$ (line 17). By repeating these steps until convergence (line 14), we ultimately obtain the optimal ride-matching mapping mechanism $C_{Z|Y}$ and minimize the corresponding utility loss.

Algorithm 7 PSO for solving $\min_{C_{Z|Y}} P(C_{Z|Y}, \delta)$

Input: $\{C_{Z|Y}^{0,1}, \cdots, C_{Z|Y}^{0,S}\}$: Positions of particle swarm, $\{V^{0,1}, \cdots, V^{0,S}\}$: Velocities of particle swarm, MNI: Maximum number of iterations, d_1, d_2: Cognitive and social acceleration coefficients

Output: $Gbest$: Best-known position of the swarm, $\{Pbest^1, \cdots, Pbest^S\}$: Best-known position of each particle, $\{V^{MNI,1}, \cdots, V^{MNI,S}\}$: Velocity of each particle

1: **for** each particle $k \in [1, S]$ **do**
2: $\quad Pbest^{0,k} \leftarrow C_{Z|Y}^{0,k}$;
3: **end for**
4: $Gbest \leftarrow Pbest^{0,1}$;
5: **for** each particle $k \in [1, S]$ **do**
6: $\quad$ **if** $P(Pbest^{0,k}, \delta) < P(Gbest, \delta)$ **then**
7: $\quad\quad Gbest \leftarrow Pbest^{0,k}$;
8: $\quad$ **end if**
9: **end for**
10: $j \leftarrow 0$;
11: **while** $j \leq MNI - 1$ **do**
12: $\quad$ **for** each particle $k \in [1, S]$ **do**
13: $\quad\quad$ Update the velocity $V^{j+1,k}$ by Eq. (4.35);
14: $\quad\quad$ Update the position $C_{Z|Y}^{j+1,k}$ by Eq. (4.36);
15: $\quad\quad$ **if** $P(C_{Z|Y}^{j+1,k}, \delta) < P(Pbest^k, \delta)$ **then**
16: $\quad\quad\quad Pbest^k \leftarrow C_{Z|Y}^{j+1,k}$;
17: $\quad\quad$ **end if**
18: $\quad\quad$ **if** $P(Pbest^k, \delta) < P(Gbest, \delta)$ **then**
19: $\quad\quad\quad Gbest \leftarrow Pbest^k$;
20: $\quad\quad$ **end if**
21: $\quad$ **end for**
22: $\quad j \leftarrow j + 1$;
23: **end while**

With a fixed penalty factor δ, we solve the problem $\min_{C_{Z|Y}} P(C_{Z|Y}, \delta)$ using PSO as described in Algorithm 7. Starting with an initial swarm composed of particles $\{C_{Z|Y}^{0,1}, \cdots, C_{Z|Y}^{0,S}\}$ and their velocities $\{V^{0,1}, \cdots, V^{0,S}\}$, we set each particle's best-known position $Pbest^k$ equal to its initial position (lines 1–3) and then identify the swarm's overall best position $Gbest$ (lines 4–9). At each iteration, the particles' velocities and positions are updated (lines 13–14) via

$$\begin{aligned} v_{m,n}^{j+1,k} =& v_{m,n}^{j,k} + d_1 \cdot r_1 \cdot (pbest_{m,n}^{k} - c_{m,n}^{j,k}) \\ &+ d_2 \cdot r_2 \cdot (gbest_{m,n}^{k} - c_{m,n}^{j,k}), \quad \forall m, n \in [1, |\mathcal{L}|], \end{aligned} \tag{4.35}$$

$$c_{m,n}^{j+1,k} = c_{m,n}^{j,k} + v_{m,n}^{j+1,k}, \quad \forall m, n \in [1, |\mathcal{L}|], \tag{4.36}$$

where r_1, r_2 are sampled from the uniform distribution $U(0, 1)$, and the indices of $pbest_{m,n}^{k} \in Pbest^{k}$, $gbest_{m,n}^{k} \in Gbest^{k}$ are defined similarly to $c_{m,n}^{j,k} \in C_{Z|Y}^{j,k}$. Subsequently, we update each particle's best-known position $Pbest^{k}$ (lines 15–17) and the swarm's overall best position $Gbest$ (lines 18–20). By repeatedly applying these updates, the l_1-penalty function is minimized, ultimately yielding the optimal swarm characterized by $Gbest$, $\{Pbest^{1}, \cdots, Pbest^{S}\}$, and $\{V^{MNI,1}, \cdots, V^{MNI,S}\}$.

4.4.3 Computational Complexity

The computational complexity of PADP-RoD comprises two main components: discretizing the spatiotemporal region (i.e., generating the adaptive supply and demand aware grid) and solving the optimization problem (O) (which involves the dynamic pricing and ride-matching sub-problems). For the grid $\mathbb{G}$, its computational complexity is $O\left(|T| \times |\mathcal{L}|\right)$, where $|T| = (ref_{per})^{ref-1} \cdot |T_o|$ and T_o is an initialized temporal grid. In particular, the CVT strategy (Algorithm 3) runs in $O(|\mathcal{L}|)$, while the computational complexity of the dichotomy strategy (lines 2–6 in Algorithm 4) is formulated as

$$\begin{aligned} &O\Big(|T_o| \cdot \log|T_o| + ref_{per} \cdot |T_o| \cdot \log(ref_{per} \cdot |T_o|) + \\ &\cdots + (ref_{per})^{ref-1} \cdot |T_o| \cdot \log((ref_{per})^{ref-1} \cdot |T_o|)\Big) \\ &\approx O\left(|T_o| \times (\log|T_o| + 1)\right). \end{aligned} \tag{4.37}$$

Since the spatial region is re-discretized at every time $\forall t_i \in T$ using the CVT approach (lines 7–11 in Algorithm 4), the overall computational complexity to obtain $\mathbb{G}$ is $O\left(|T_o| \times (\log|T_o| + 1) + |T| \times |\mathcal{L}|\right) \approx O(|T| \times |\mathcal{L}|)$.

Regarding the sub-problems, solving (O_d) and (O_r) incurs computational complexities of $O(|\mathcal{L}| \times |\mathcal{L}|)$ and $O(S \times |\mathcal{L}|)$, respectively. Notably, the computational complexities for (O_r) is largely due to Algorithm 7 (line 16 in Algorithm 6), although the ride-matching sub-problem generally converges within only a few iterations. Since Algorithm 7 itself is $O(S \times |\mathcal{L}|)$, obtaining the mapping mechanism

$C_{Z|Y}$ also requires $O(S \times |\mathcal{L}|)$. In summary, the total complexity of PADP-RoD is $O((|T| + |\mathcal{L}| + S) \times |\mathcal{L}|)$, which simplifies to $O(|\mathcal{L}| \times |\mathcal{L}|)$ when $|\mathcal{L}| \gg |T| + S$. Moreover, the local mapping mechanism $\mathcal{A}_{Z|X}$ can be computed in advance, allowing efficient local perturbation of users' real locations.

4.5 Privacy Analysis

In this part, we prove that PADP-RoD satisfies ε_d-DP and ε_d-identifiability, and their detailed theorems are given below.

Theorem 4.1 *Let $\varepsilon_d \geq 0$. Given the dynamic pricing mapping mechanism $\mathcal{B}_{Y|X}$ satisfies ε_d-DP. We have PADP-RoD satisfies ε_d-DP for any ride-matching mapping mechanism $C_{Z|Y}$.*

Proof Because $\mathcal{A}_{Z|X}(z \mid x) = \sum_{y \in \mathcal{L}} C_{Z|Y}(z \mid y) \cdot \mathcal{B}_{Y|X}(y \mid x)$, we have

$$\frac{\mathcal{A}_{Z|X}(z \mid x)}{\mathcal{A}_{Z|X}(z \mid x')} = \frac{\sum_{y \in \mathcal{L}} C_{Z|Y}(z \mid y) \cdot \mathcal{B}_{Y|X}(y \mid x)}{\sum_{y \in \mathcal{L}} C_{Z|Y}(z \mid y) \cdot \mathcal{B}_{Y|X}(y \mid x')}, \tag{4.38}$$

where locations x and x' are neighbors. Since the mechanism $\mathcal{B}_{Y|X}$ satisfies ε_d-DP, we have

$$\frac{\mathcal{B}_{Y|X}(y \mid x)}{\mathcal{B}_{Y|X}(y \mid x')} \leq e^{\varepsilon_d}, \ \forall x, x' : x \sim x', y \in \mathcal{L}. \tag{4.39}$$

Thus, we have

$$\frac{\sum_{y \in \mathcal{L}} C_{Z|Y}(z \mid y) \cdot \mathcal{B}_{Y|X}(y \mid x)}{\sum_{y \in \mathcal{L}} C_{Z|Y}(z \mid y) \cdot \mathcal{B}_{Y|X}(y \mid x')} \tag{4.40}$$

$$\leq \frac{\sum_{y \in \mathcal{L}} C_{Z|Y}(z \mid y) \cdot e^{\varepsilon_d} \cdot \mathcal{B}_{Y|X}(y \mid x')}{\sum_{y \in \mathcal{L}} C_{Z|Y}(z \mid y) \cdot \mathcal{B}_{Y|X}(y \mid x')}, \tag{4.41}$$

$$= e^{\varepsilon_d}. \tag{4.42}$$

In summary, we have

$$\frac{\mathcal{A}_{Z|X}(z \mid x)}{\mathcal{A}_{Z|X}(z \mid x')} \leq e^{\varepsilon_d}, \ \forall x, x' : x \sim x', z \in \mathcal{L}. \tag{4.43}$$

That is, the local mapping mechanism $\mathcal{A}_{Z|X}$ satisfies ε_d-DP. □

Theorem 4.2 *Let $\varepsilon_d \geq 0$. Given the ride-matching mapping mechanism $C_{Z|Y}$ satisfies $\varepsilon_d/2$-DP. We have PADP-RoD satisfies ε_d-DP for any dynamic pricing mapping mechanism $\mathcal{B}_{Y|X}$.*

Proof Because $\mathcal{A}_{Z|X}(z \mid x) = \sum_{y\in\mathcal{L}} C_{Z|Y}(z \mid y) \cdot \mathcal{B}_{Y|X}(y \mid x)$, we have

$$\frac{\mathcal{A}_{Z|X}(z \mid x)}{\mathcal{A}_{Z|X}(z \mid x')} = \frac{\sum_{y\in\mathcal{L}} C_{Z|Y}(z \mid y) \cdot \mathcal{B}_{Y|X}(y \mid x)}{\sum_{y\in\mathcal{L}} C_{Z|Y}(z \mid y) \cdot \mathcal{B}_{Y|X}(y \mid x')}, \tag{4.44}$$

where locations x and x' are neighbors. Since the mechanism $C_{Z|Y}$ satisfies $\varepsilon_d/2$-DP, we have

$$C_{Z|Y}(z \mid y) \leq e^{\varepsilon_d/2} \cdot C_{Z|Y}(z \mid y'), \forall y, y' : y \sim y', z \in \mathcal{L}, \tag{4.45}$$

$$e^{-\varepsilon_d/2} \cdot C_{Z|Y}(z \mid y') \leq C_{Z|Y}(z \mid y), \forall y, y' : y \sim y', z \in \mathcal{L}. \tag{4.46}$$

According to Eqs. (4.45) and (4.46), we have

$$\frac{\mathcal{A}_{Z|X}(z \mid x)}{\mathcal{A}_{Z|X}(z \mid x')} \leq \frac{\sum_{y\in\mathcal{L}} e^{\varepsilon_d/2} \cdot C_{Z|Y}(z \mid y') \cdot \mathcal{B}_{Y|X}(y \mid x)}{\sum_{y\in\mathcal{L}} e^{-\varepsilon_d/2} \cdot C_{Z|Y}(z \mid y') \cdot \mathcal{B}_{Y|X}(y \mid x')}, \tag{4.47}$$

$$= e^{\varepsilon_d} \times \frac{\sum_{y\in\mathcal{L}} \mathcal{B}_{Y|X}(y \mid x)}{\sum_{y\in\mathcal{L}} \mathcal{B}_{Y|X}(y \mid x')}, \tag{4.48}$$

$$\overset{(a)}{=} e^{\varepsilon_d}, \tag{4.49}$$

where (a) follows the fact that $\sum_{y\in\mathcal{L}} \mathcal{B}_{Y|X}(y \mid x) = 1 = \mathcal{B}_{Y|X}(y \mid x')$. In summary, the local mapping mechanism $\mathcal{A}_{Z|X}$ satisfies ε_d-DP. □

Theorem 4.3 *The proposed PADP-RoD satisfies both ε_d-DP and ε_d-identifiability.*

Proof The DP constraints of sub-problems (O_d) and (O_r), i.e., constraints (4.15b) and (4.25b), guarantee that the dynamic pricing mapping mechanism $\mathcal{B}_{Y|X}$ and the ride-matching mapping mechanism $C_{Z|Y}$ satisfies ε_d-DP and $\varepsilon_d/2$-DP, respectively. Thus, the local mapping mechanism $\mathcal{A}_{Z|X}$ satisfies ε_d-DP according to the Theorems 4.1 and 4.2. Besides, it is known from [31, Theorem 1] that if the mechanism $\mathcal{A}_{Z|X}$ satisfies ε_d-DP then it also satisfies ε_d-identifiability. In summary, the mechanism $\mathcal{A}_{Z|X}$ satisfies both ε_d-DP and ε_d-identifiability. □

4.6 Performance Evaluation

In this section, we evaluate the utility (i.e., dynamic pricing utility and ride-matching utility) and location privacy of PADP-RoD on the real-world event-log dataset from an RoD service.

4.6.1 Simulation Settings

All experiments are conducted using the same dataset and implementation environment. The simulations are implemented in Python and performed on a laptop with an Intel Core i7-8750H 2.20 GHz processor and 16GB RAM.

Dataset Our event-log data originates from Shenzhou UCar,[1] one of China's leading RoD service providers. The dataset comprises 14,587,353 records collected over four months (August-November 2016) in Beijing, with all data anonymized. Each record represents a single event and includes details such as Event_time, Event_boarding_location (longitude and latitude), Event_arriving_location (longitude and latitude), Estimated_fare, and Price_multiplier. Notably, the price multiplier is restricted between 1.0 and 1.6, with possible values of 1.0, 1.1, 1.2, 1.3, 1.4, 1.5, and 1.6. Table 4.2 presents two sample entries from the event-log dataset. Additionally, we collected GPS data of drivers during the same period from Shenzhou UCar. Each GPS record contains the record time, driver ID, longitude, and latitude. Table 4.3 displays two example GPS records.

Baselines In our evaluation, PADP-RoD is compared against three notable approaches from the literature: Dif-Dis [16], MAL-Geo-I [35], and Laplace [27]. Below is a brief introduction to each of the methods.

- *Dif-Dis [16].* This method delivers high-quality ride-matching for drivers while ensuring location privacy through geo-indistinguishability [27] combined with distortion privacy. Specifically, Dif-Dis computes an average travel distance based on the rider's real location, the driver's perturbed location, and the employed mapping mechanism. The problem is then cast as a single-objective optimization problem that minimizes this average distance under both geo-indistinguishability and distortion privacy constraints.
- *MAL-Geo-I [35].* MAL-Geo-I is designed to minimize the average loss while delivering location privacy protection that complies with the principles of geo-indistinguishability [27].

Table 4.2 Sample entries in our event-log dataset

Event_time	Boarding_lon	Boarding_lat	Arriving_lon	Arriving_lat	Fare	Price_multiplier
01/11/16 09:09	116.4658203	39.8971405	116.4675064	39.9571304	44	1.0
30/11/16 14:58	116.1534729	39.8046532	116.3790894	39.8648796	344	1.6

Table 4.3 Sample records in our GPS dataset

Record_time	Driver_ID	Lon	Lat
01/11/16 00:04	NV7H87	116.733986	40.20458
01/11/16 23:59	QX2F69	116.48728	39.9444

[1] http://www.10101111.com.

- *Laplace [27].* This approach perturbs the real location by adding Laplace noise, thereby rigorously satisfying geo-indistinguishability [27]. The obfuscation probabilities are defined as follows:

$$\mathcal{A}_{Z|X}(z \mid x) \propto e^{-\varepsilon_d \cdot \frac{d(x,z)}{D(\mathcal{L})}}, \tag{4.50}$$

where $D(\mathcal{L})$ is the maximum geo-distance between any two locations in $\mathcal{L}$.

Metrics We assess PADP-RoD based on four key aspects: the adaptive supply and demand aware grid, the dynamic pricing utility, the ride-matching utility, and the location privacy.

- *Adaptive Supply and Demand Aware Grid.* To measure grid quality, we employ the Average Price Multiplier Error (APME), defined as

$$\text{APME} = \frac{\sqrt{\sum_{l_i, t_j \in D_{real}} \left(m(l_i, t_j) - \mathbb{E}_m(\mathbb{G}(l_i, t_j))\right)^2}}{|D_{real}|}, \tag{4.51}$$

where $m(l_i, t_j)$ is the price multiplier at the point (l_i, t_j), $\mathbb{G}(l_i, t_j)$ identifies the corresponding grid cell (e.g., cell C_n^k), and $\mathbb{E}_m(*)$ computes the average multiplier within that cell. A lower APME indicates a higher-quality grid.
- *Dynamic Pricing Utility.* Since the price multiplier is driven by the prevailing supply-demand balance, we quantify dynamic pricing utility using the Average Supply-Demand Relationship Error (ASDRE), defined as follows.

Definition 4.12 (Average Supply-Demand Relationship Error (ASDRE)) Let $S(x,t)/D(x,t)$ denote the actual numbers of drivers/riders in a circular area centered at x during period t, and let $S^*(x,t)/D^*(x,t)$ represent their perturbed numbers. The error is the difference between the true imbalance $S(x,t)-D(x,t)$ and the perturbed imbalance $S^*(x,t) - D^*(x,t)$. The average supply-demand relationship error is formulated as

$$\text{ASDRE} = \frac{1}{|T| \cdot |\mathcal{L}|} \cdot \sum_{t \in T} \sum_{x \in \mathcal{L}} |(S(x,t) - D(x,t)) - \left(S^*(x,t) - D^*(x,t)\right)|. \tag{4.52}$$

Definition 4.12 presents the lower ASDRE, the better the obtained dynamic pricing utility.

- *Ride-Matching Utility.* We measure the efficiency of ride-matching using the RMUM metric (as detailed in Definition 4.10), and a smaller RMUM value signifies superior ride-matching utility.
- *Location Privacy.* Location privacy is measured using the Average Bayesian Inference Error (ABIE), a standard metric for evaluating resistance to Bayesian inference attacks, defined as follows.

Definition 4.13 (Average Bayesian Inference Error (ABIE)) We define the Bayesian inference error as the geo-distance between the inferred location $\widehat{x}$ and real location x of the observed location z (detailed in definition 4.1). Then, the average Bayesian inference error can be formulated as

$$\text{ABIE} = \frac{1}{D_{real}} \sum_{z \in D_{real}} d(\widehat{x}, x). \tag{4.53}$$

Obviously, a higher ABIE value indicates stronger location privacy protection.

Parameter Settings As outlined in Sect. 4.4, we consider several crucial parameters: the differential privacy budget ε_d, the number of generators g_n, and the refinement number ref. For our evaluation, we vary ε_d over $\{0.1, 0.3, 0.5, 0.7, 0.9\}$, g_n over $\{50, 100, 150, 200, 250, 300\}$, and ref over $\{6, 12, 18, 24, 30\}$. In addition, several empirical parameters within PADP-RoD are set as follows. In Algorithm 3, the convergence threshold is $\eta = 0.01$, and the percentage of cells that need to be refined $ref_{per} = 0.1$. Algorithm 5 employs a convergence threshold $\theta_1 = 0.01$. In Algorithm 6, following [33], the settings are: the threshold for convergence $\theta_2 = 0.01$, the penalty factor $\delta = 10$, the penalty factor growth coefficient $\rho = 10$, and the maximum dynamic pricing utility loss that allows damage $\Phi = 0.1 \cdot \arg\min Pri(\mathcal{B}_{Y|X})$. Algorithm 7, based on [34], is configured with the swarm size $S = 100$, the maximum number of iterations $MNI = 1000$, and the cognitive and social acceleration coefficients $d_1 = 2, d_2 = 2$. Finally, the target area Ω is partitioned into a grid of 114×90 cells, each measuring 500×500 m, with the location set $\mathcal{L}$ defined by the centers of these cells.

4.6.2 *Simulation Results*

We evaluate the effectiveness of PADP-RoD from four perspectives: adaptive supply and demand aware grid, dynamic pricing utility, ride-matching utility, and privacy protection.

Quality of the Adaptive Supply and Demand Aware Grid The proposed adaptive grid is benchmarked against two existing methods, the uniform grid [29] and the two-step grid [17]. The uniform grid method partitions the spatial region Ω and temporal region T into cells of equal size, while the two-step grid initially uses uniform divisions and then further splits each cell to capture variations in the price multiplier. As depicted in Fig. 4.6, our approach yields a significantly lower Average Price Multiplier Error (APME) than the other two methods. For example, when the number of generators $g_n = 50$, the APME values in the spatial dimension are 0.07 for our grid, compared to 0.14 for the uniform grid and 0.09 for the two-step grid. This demonstrates that our grid achieves lower APME with fewer cells, primarily due to its adaptive cell sizing based on price multiplier elasticity.

Dynamic Pricing Utility Guarantee Figure 4.7a presents the dynamic pricing performance of PADP-RoD compared to baseline methods using the Shenzhou

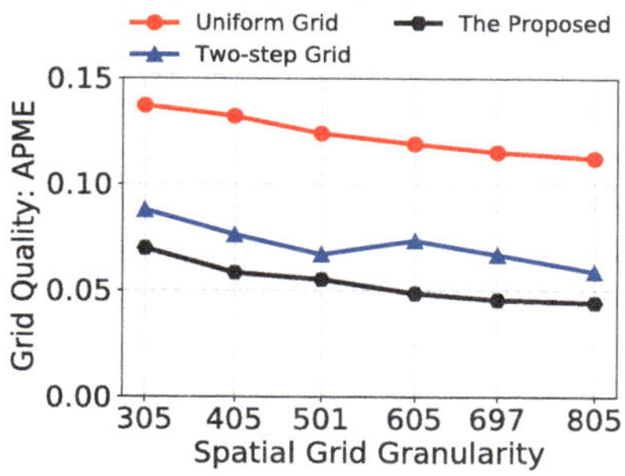

(a) Impact of g_n on grid quality

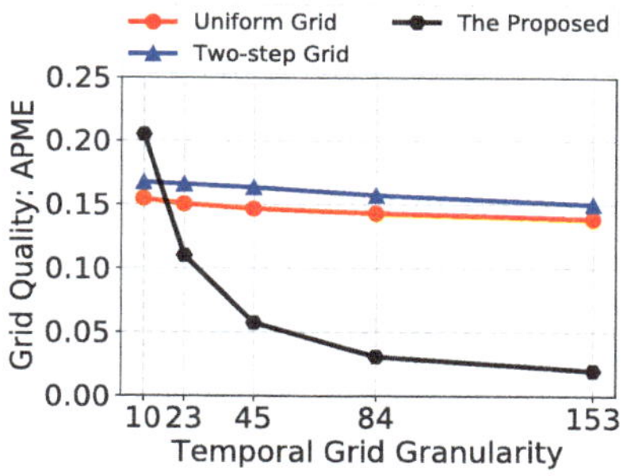

(b) Impact of ref on grid quality

Fig. 4.6 The quality of different grids in a real-world event-log dataset varies with the number of cells. Spatial grid granularity (x axis) represents the number of spatial cells, and g_n = {50, 100, 150, 200, 250, 300} means spatial grid granularity: {305, 405, 501, 605, 697, 805}. Similarly, temporal grid granularity (x axis) represents the number of temporal cells, and ref = {6, 12, 18, 24, 30} means temporal grid granularity: {10, 23, 45, 84, 153}. (**a**) Average price multiplier error varying with g_n. (**b**) Average price multiplier error varying with ref

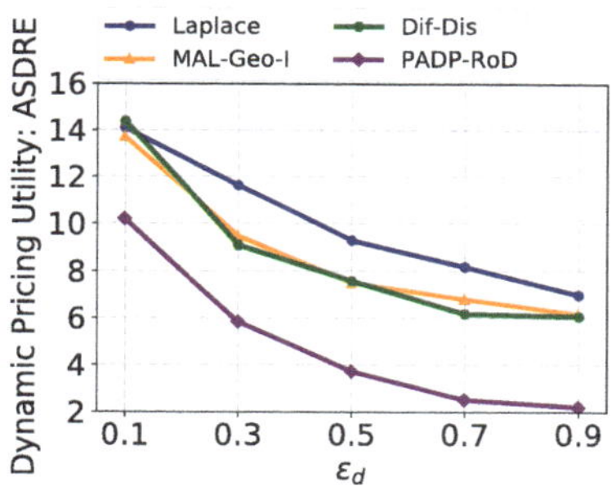

(a) Impact of ε_d on dynamic pricing utility

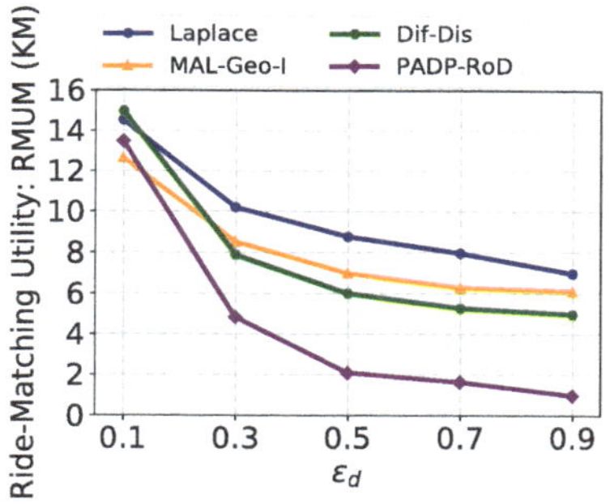

(b) Impact of ε_d on ride-matching utility

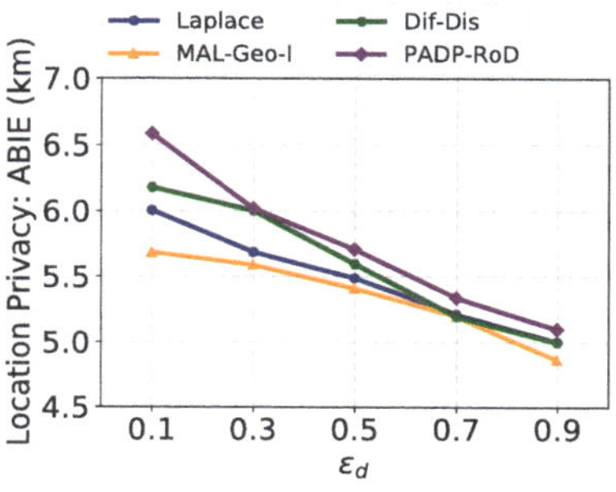

(c) Impact of ε_d on location privacy

Fig. 4.7 (**a**) Dynamic pricing utility loss of different location-privacy preserving mechanisms. (**b**) Ride-Matching utility loss of different location-privacy preserving mechanisms. (**c**) Effectiveness of different location privacy-preserving mechanisms defends against the Bayesian inference attack

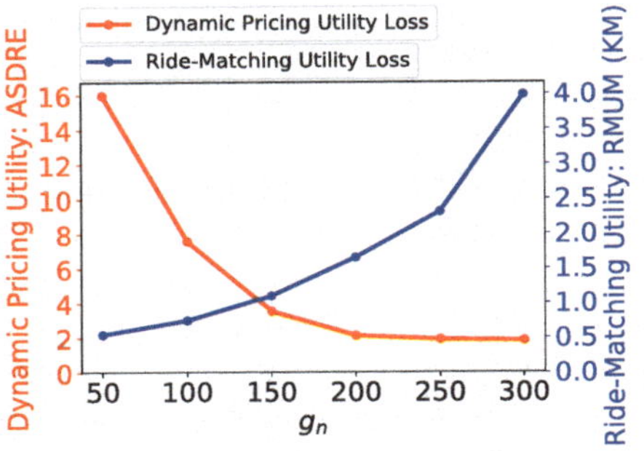

(a) Impact of ε_d on dynamic pricing utility

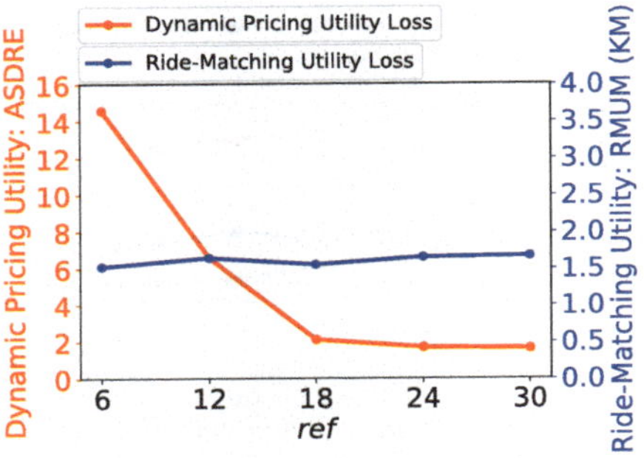

(b) Impact of ε_d on ride-matching utility

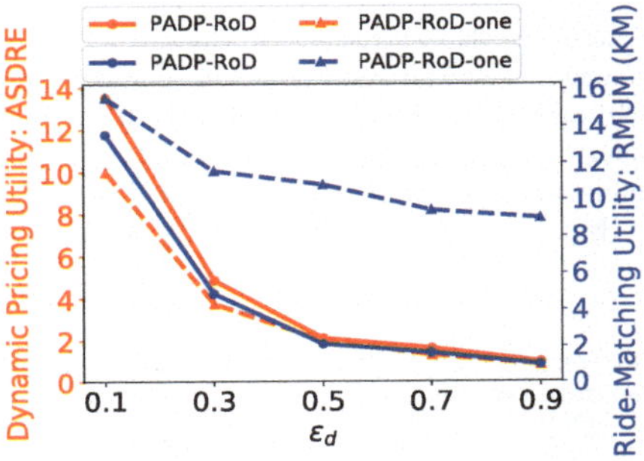

(c) Impact of ε_d on location privacy

Fig. 4.8 Utility loss (the red line represents the dynamic pricing utility loss and the blue line represents the ride-matching utility loss) in a real-word event-log dataset varying with the number of generators g_n, the refinement number ref, and the differential privacy budget ε_d. The dynamic pricing utility loss is measured by the Average Supply-Demand Relationship Error (ASDRE) and the ride-matching utility loss is measured by the Ride-Matching Utility Metric (RMUM). (**a**) Utility loss varies with g_n. (**b**) Utility loss varies with ref. (**c**) Utility loss varies with ε_d, where the comparison of PADP-RoD and PADP-RoD-one (consists only of dynamic pricing mapping mechanism $\mathcal{B}_{Y|X}$) is used to evaluate whether the mapping mechanism $\mathcal{B}_{Y|X}$ can preserve the ride-matching utility and whether the mapping mechanism $C_{Z|Y}$ can damage the dynamic pricing utility

UCar event-log dataset. It is evident that PADP-RoD achieves a much lower Average Supply-Demand Relationship Error (ASDRE) than Dif-Dis, MAL-Geo-I, and Laplace. For example, with a privacy budget of $\varepsilon_d = 0.9$, the ASDRE values are 2.22 for PADP-RoD, versus 7.01, 6.20, and 6.09 for Laplace, Dif-Dis, and MAL-Geo-I, respectively. This advantage arises because the baseline methods tend to overlook the adverse effects of noise on the quality of the dynamic pricing service. Moreover, Fig. 4.8a, b (red solid line) explore the impact of varying the number of generators g_n and the refinement number ref on dynamic pricing utility. We observe that as g_n increases from 50 to 200, the ASDRE decreases rapidly. A similar trend is seen with ref, increasing it from 6 to 18 results in a sharp decline in ASDRE, since a larger number of cells better captures the dynamic nature of the price multiplier.

However, once the cell count is sufficiently high, further improvements become marginal (as detailed in Fig. 4.6). Finally, Fig. 4.8c shows that the ASDRE for PADP-RoD is nearly identical to that of PADP-RoD-one (which employs only the dynamic pricing mapping mechanism $\mathcal{B}_{Y|X}$). This confirms that the additional ride-matching mapping mechanism $C_{Z|Y}$ does not significantly compromise the dynamic pricing service quality.

Ride-Matching Utility Guarantee Figure 4.7b illustrates the trade-off between ride-matching utility and location privacy for PADP-RoD compared to other methods in our dataset. Notably, PADP-RoD achieves markedly better ride-matching performance, as indicated by a substantially lower Ride-Matching Utility Metric (RMUM). For instance, when the privacy budget is $\varepsilon_d = 0.9$, PADP-RoD records an RMUM of 1.00, whereas Laplace, Dif-Dis, and MAL-Geo-I yield values of 7.00, 6.17, and 5.00, respectively. This advantage stems from the fact that baseline methods, such as MAL-Geo-I and Laplace, overlook the service quality degradation due to noise; MAL-Geo-I, in particular, estimates geo-distances using expected travel distances without minimizing the corresponding errors. Furthermore, Fig. 4.8a, b (blue solid line) explore the effects of the number of generators g_n and the refinement number ref on ride-matching utility. The results reveal that as g_n increases from 50 to 200, RMUM rises gradually before increasing sharply with further increments. In contrast, changes in ref do not significantly affect the ride-matching utility. This suggests that an excessively high g_n, which results in overly small grid cells, makes it challenging to find suitable perturbed locations, thereby compromising the utility. Finally, Fig. 4.8c (blue solid line versus blue dashed line) shows that PADP-RoD outperforms PADP-RoD-one (which only employs the dynamic pricing mapping $\mathcal{B}_{Y|X}$ by achieving a significantly lower RMUM. This confirms that the additional ride-matching mapping mechanism $C_{Z|Y}$ is essential for delivering high-quality ride-matching service.

Effectiveness of Privacy Protection We assess the location privacy provided by PADP-RoD using our dataset. Figure 4.7c illustrates that PADP-RoD, along with Dif-Dis, MAL-Geo-I, and Laplace, effectively counteracts Bayesian inference attacks. For instance, with a privacy budget of $\varepsilon_d = 0.9$, the ABIE scores are 5.10 for PADP-RoD, 5.00 for Laplace, 4.87 for Dif-Dis, and 5.01 for MAL-Geo-I. This similarity in performance stems from all methods adhering to ε_d-differential privacy or its generalized form ε_d-Geo-indistinguishability. Additionally, Fig. 4.9 (with the magenta line representing PADP-RoD) examines how varying the differential privacy budget ε_d, the number of generators g_n, and the refinement number ref on the location privacy. The data reveal that the Average Bayesian Inference Error (ABIE) decreases as ε_d increases, indicating that a higher privacy budget leads to weaker privacy protection. In contrast, g_n and ref mainly influence dynamic pricing and ride-matching utilities, leaving ABIE largely unaffected. To further validate our theoretical findings (Theorems 4.1, 4.2, and 4.3), we introduce two variants: PADP-RoD-AOne and PADP-RoD-ATwo. Specifically, PADP-RoD-AOne combines an optimal dynamic pricing mapping $\mathcal{B}_{Y|X}$ with a non-optimized ride-matching map-

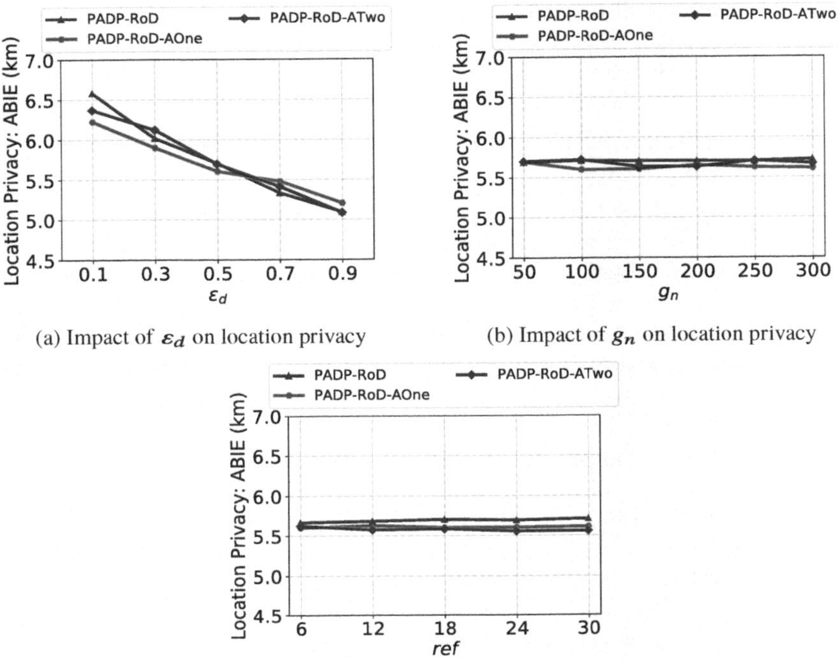

(a) Impact of ε_d on location privacy

(b) Impact of g_n on location privacy

(c) Impact of ref on location privacy

Fig. 4.9 Average bayesian inference error (ABIE) in a real-word event-log dataset varying with the differential privacy budget ε_d, the number of generators g_n, and the refinement number ref. PADP-RoD-AOne consists of an optimal stage-one privacy mapping mechanism $\mathcal{B}_{Y|X}$ and an arbitrary stage-two privacy mapping mechanism $C_{Z|Y}$. Similarly, PADP-RoD-ATwo consists of an arbitrary $\mathcal{B}_{Y|X}$ and an optimal $C_{Z|Y}$. Additionally, our propose PADP-RoD consists of the optimal $\mathcal{B}_{Y|X}$ and $C_{Z|Y}$. (**a**) ABIE varies with ε_d. (**b**) ABIE varies with g_n. (**c**) ABIE varies with ref

ping $C_{Z|Y}$, while PADP-RoD-ATwo uses a non-optimized dynamic pricing mapping alongside an optimal ride-matching mapping. As shown in Fig. 4.9, both variants and the fully optimized PADP-RoD offer comparable privacy protection against Bayesian inference attacks, corroborating our privacy analysis results.

4.7 Summary and Further Reading

In this chapter, we introduced PADP-RoD, a pricing-aware differentially private framework designed for ride-on-demand services. This framework ensures high-quality dynamic pricing and ride-matching services while protecting users' location data privacy. Additionally, the proposed utility metrics for dynamic pricing and ride-matching are broadly applicable, effectively measuring quality loss due to perturbation in sporadic location privacy scenarios. Notably, PADP-RoD can be extended to other location-based applications, such as mobile crowdsensing, to enhance service quality while protecting location privacy. For the future work, we plan to explore a continuous location privacy protection mechanism that prevents users' location trace data leaks while providing high-quality location-based services.

References

1. Guo, S., Chen, C., Wang, J., Liu, Y., Xu, K., Zhang, D., Chiu, D.M.: A simple but quantifiable approach to dynamic price prediction in ride-on-demand services leveraging multi-source urban data. Proc. ACM Interact. Mobile Wearable Ubiq. Technol. **2**(3), 1–24 (2018)
2. Guo, S., Chen, C., Liu, Y., Xu, K., Chiu, D.M.: Modelling passengers' reaction to dynamic prices in ride-on-demand services: a search for the best fare. Proc. ACM Interact. Mobile Wearable Ubiq. Technol. **1**(4), 1–23 (2018)
3. Guo, S., Chen, C., Wang, J., Liu, Y., Xu, K., Yu, Z., Zhang, D., Chiu, D.M.: Rod-revenue: seeking strategies analysis and revenue prediction in ride-on-demand service using multi-source urban data. IEEE Trans. Mobile Comput. **19**(9), 2202–2220 (2020)
4. Guo, S., Chen, C., Wang, J., Ding, Y., Liu, Y., Xu, K., Yu, Z., Zhang, D.: A force-directed approach to seeking route recommendation in ride-on-demand service using multi-source urban data. IEEE Trans. Mobile Comput. **21**(6), 1909–1926 (2022)
5. Chen, M.K.: Dynamic pricing in a labor market: surge pricing and flexible work on the uber platform. In: Proceedings of ACM Conference on Economics and Computation, pp. 1–19 (2016)
6. Huang, J., Huang, L., Liu, M., Li, H., Tan, Q., Ma, X., Cui, J., Huang, D.S.: Deep reinforcement learning-based trajectory pricing on ride-hailing platforms. ACM Trans. Intell. Syst. Technol. **13**(3), 1–19 (2022)
7. Shao, S., Mittal, A., Twumasi-Boakye, R., Gupta, A.: Fair pricing of ridehailing services with asymmetric demand and travel time. IEEE Trans. Control Netw. Syst. **9**(2), 670–681 (2022)
8. Jiang, H., Li, J., Zhao, P., Zeng, F., Xiao, Z., Iyengar, A.: Location privacy-preserving mechanisms in location-based services: a comprehensive survey. ACM Comput. Surv. **54**(1), 1–36 (2021)
9. de Montjoye, Y.A., Hidalgo, C.A., Verleysen, M., Blondel, V.D.: Unique in the crowd: the privacy bounds of human mobility. Sci. Rep. **3**(1), 2045–2322 (2013)
10. Zheng, Z., Li, Z., Jiang, H., Zhang, L.Y., Tu, D.: Semantic-aware privacy-preserving online location trajectory data sharing. IEEE Trans. Inf. Forensics Secur. **17**, 2256–2271 (2022)
11. Zheng, Z., Li, Z., Li, J., Jiang, H., Li, T., Guo, B.: Utility-aware and privacy-preserving trajectory synthesis model that resists social relationship privacy attacks. ACM Trans. Intel. Syst. Technol. **13**(3), 1–28 (2022)
12. Luo, Y., Jia, X., Fu, S., Xu, M.: Pride: privacy-preserving ride matching over road networks for online ride-hailing service. IEEE Trans. Inf. Forensics Secur. **14**(7), 1791–1802 (2019)
13. Huang, J., Luo, Y., Fu, S., Xu, M., Hu, B.: Pride: privacy-preserving online ride hailing matching system with prediction. IEEE Trans. Veh. Technol. **70**(8), 7413–7425 (2021)
14. Wang, L., Yang, D., Han, X., Wang, T., Zhang, D., Ma, X.: Location privacy-preserving task allocation for mobile crowdsensing with differential geo-obfuscation. In: Proceedings of International Conference on World Wide Web, pp. 627–636 (2017)
15. Wang, L., Zhang, D., Yang, D., Lim, B.Y., Han, X., Ma, X.: Sparse mobile crowdsensing with differential and distortion location privacy. IEEE Trans. Inf. Forensics Secur. **15**, 2735–2749 (2020)

16. Wang, L., Yang, D., Han, X., Zhang, D., Ma, X.: Mobile crowdsourcing task allocation with differential-and-distortion geo-obfuscation. IEEE Trans. Depend. Secur. Comput. **18**(2), 967–981 (2021)
17. Gursoy, M.E., Liu, L., Truex, S., Yu, L., Wei, W.: Utility-aware synthesis of differentially private and attack-resilient location traces. In: Proceedings of ACM SIGSAC Conference on Computer and Communications Security, pp. 196–211 (2018)
18. Bindschaedler, V., Shokri, R.: Synthesizing plausible privacy-preserving location traces. In: Proceedings of IEEE Symposium on Security and Privacy, pp. 546–563 (2016)
19. Zheng, Z., Li, Z., Huang, C., Long, S., Li, M., Shen, X.: Data poisoning attacks and defenses to ldp-based privacy-preserving crowdsensing. IEEE Trans. Depend. Secur. Comput. **21**(5), 4861–4878 (2024)
20. Li, Z., Zheng, Z., Guo, S., Guo, B., Xiao, F., Ren, K.: Disguised as privacy: data poisoning attacks against differentially private crowdsensing systems. IEEE Trans. Mobile Comput. **22**(9), 5155–5169 (2023)
21. Duchi, J.C., Jordan, M.I., Wainwright, M.J.: Local privacy and statistical minimax rates. In: Proceedings of IEEE Annual Symposium on Foundations of Computer Science, pp. 429–438 (2013)
22. BBC: Didi shares fall on reports china is planning penalties (Jul. 2021). https://www.bbc.com/news/business-57938212
23. Shokri, R., Theodorakopoulos, G., Troncoso, C., Hubaux, J.P., Le Boudec, J.Y.: Protecting location privacy: optimal strategy against localization attacks. In: Proceedings of ACM Conference on Computer and Communications Security, pp. 617–627 (2012)
24. Oya, S., Troncoso, C., Pérez-González, F.: Rethinking location privacy for unknown mobility behaviors. In: Proceedings of IEEE European Symposium on Security and Privacy, pp. 416–431 (2019)
25. Dwork, C., McSherry, F., Nissim, K., Smith, A.: Calibrating noise to sensitivity in private data analysis. J. Privacy Confidential. **7**(3), 17–51 (2017)
26. Wang, W., Ying, L., Zhang, J.: On the relation between identifiability, differential privacy, and mutual-information privacy. IEEE Trans. Inf. Theory **62**(9), 5018–5029 (2016)
27. Andrés, M.E., Bordenabe, N.E., Chatzikokolakis, K., Palamidessi, C.: Geo-indistinguishability: differential privacy for location-based systems. In: Proceedings of ACM SIGSAC Conference on Computer and Communications Security, pp. 901–914 (2013)
28. Du, Q., Faber, V., Gunzburger, M.: Centroidal voronoi tessellations: applications and algorithms. SIAM Rev. **41**(4), 637–676 (1999)
29. Acs, G., Castelluccia, C.: A case study: privacy preserving release of spatio-temporal density in paris. In: Proceedings of ACM SIGKDD International Conference on Knowledge Discovery and Data Mining, pp. 1679–1688 (2014)
30. Zhao, Q., Zuo, C., Pellegrino, G., Lin, Z.: Geo-locating drivers: a study of sensitive data leakage in ride-hailing services. In: Proceedings of Annual Network and Distributed System Security Symposium, pp. 1–15 (2019)
31. Sun, M., Tay, W.P.: On the relationship between inference and data privacy in decentralized IoT networks. IEEE Trans. Inf. Forensics Secur. **15**, 852–866 (2020)
32. Gramlich, G., Hettich, R., Sachs, E.W.: Local convergence of SQP methods in semi-infinite programming. SIAM J. Optim. **5**(3), 641–658 (1995)
33. Price, C., Coope, I.: An exact penalty function algorithm for semi-infinite programmes. BIT Numer. Math. **30**, 723–734 (1990)
34. Poli, R., Kennedy, J., Blackwell, T.: Particle swarm optimization. Swarm Intel. **1**(1), 33–57 (2007)
35. Oya, S., Troncoso, C., Pérez-González, F.: Back to the drawing board: revisiting the design of optimal location privacy-preserving mechanisms. In: Proceedings of ACM SIGSAC Conference on Computer and Communications Security, pp. 1959–1972 (2017)

Chapter 5
Data Poisoning Attacks and Defenses to LDP-Based Crowdsensing

5.1 Introduction

Crowdsensing has become a prevalent method for gathering information in fields like healthcare and transportation. In these systems, mobile devices such as wearables and connected vehicles are enlisted by a crowdsensing server to function as intelligent sensors, often called workers, that collect diverse environmental data, including temperature and humidity [1–4]. This model stands in contrast to conventional sensor networks that depend on fixed installations, as it taps into the mobility of the crowd, reducing deployment costs while offering broader coverage and improved accuracy.

However, two major challenges hinder the further development of crowdsensing. First, due to the varying sensing capabilities of different workers, the quality of collected data can be inconsistent, leading to unreliable or conflicting information [5]. In addition, some participants might deliberately manipulate their reports through data poisoning attacks to gain an unfair advantage [6–8]. For example, in smart transportation, dishonest drivers could falsify information about traffic jams or accidents to divert other vehicles and secure less congested routes for themselves. Second, since these tasks are performed by individual users, the data often contains sensitive personal details such as location and health information [9–16], which raises serious privacy issues and may deter participation.

Many state-of-the-art works [5, 17–23] have been proposed to tackle these two significant issues. In order to determine reliable information from conflicting sensory reports in crowdsensing, researchers typically rely on truth discovery techniques [5, 17–19]. Many of these techniques utilize weighted aggregation algorithms that assess the trustworthiness of workers based on their agreement with the majority. Specifically, workers whose data consistently deviates from the consensus receive lower weights, thereby reducing the influence of potentially corrupted reports. In parallel, the adoption of Local Differential Privacy (LDP) [20–22] has grown as a means to protect sensing data privacy in crowdsensing systems. Under

Z. Zheng et al., *Perturbation Based Privacy in Crowdsensing*, Wireless Networks,
https://doi.org/10.1007/978-3-031-95052-0_5

LDP, individual workers add noise to their data before uploading, with the level of noise perturbation governed by a predefined privacy budget. A smaller privacy budget results in higher noise levels, thereby offering stronger privacy protection.

While both truth discovery and LDP have proven useful in their own right, their combined application inadvertently creates vulnerabilities. The noise introduced by LDP can make the overall aggregation process highly sensitive to even minor modifications in the data. This sensitivity enables malicious participants to subtly manipulate the data distribution and ultimately distort the inferred truth. Previous research [24–26] has revealed that privacy-preserving mechanisms for tasks such as identifying heavy hitters and constructing histograms under LDP are particularly prone to data poisoning. These studies generally assume that all workers contribute equally to the aggregation process, and the defenses they propose tend to be effective only against simple and static attack strategies. As a result, there is a critical need for integrated studies that address both attack methodologies and their corresponding defenses.

Our work diverges from prior research by concentrating on more sophisticated poisoning strategies aimed at LDP-based privacy-preserving crowdsensing systems. In these settings, the perturbation introduced by LDP is treated as a calibrated noise, which allows attackers to blend their malicious modifications into the expected data variation. Two primary challenges must be overcome to execute such attacks. The first challenge is *concealing malicious behavior within random noise introduced by LDP,* since the exact noise distribution is not directly observable. The second challenge is *maximizing the disruptive impact on the aggregated truth while maintaining a stealthy profile.* It is noted that the attack behavior can remain stealthy if the malicious behavior and LDP perturbation are indistinguishable, but it does not degrade the trustworthiness of inferred truth information. Thus, it is not easy for attackers to achieve a significant deviation from the outcomes without revealing their presence.

To address these challenges, we propose an optimal data poisoning attack framework specifically designed for LDP-based privacy-preserving crowdsensing. The attack is formulated as an optimization problem, maximizing the deviation from the outputs of crowdsensing while operating under the constraints imposed by the privacy-preserving truth discovery methods. Given that the constraint itself is formulated as an optimization problem, the formulated attack model becomes bi-level and NP-hard. We tackle this complexity by decomposing the problem into upper-level and lower-level sub-problems, which are solved iteratively using the augmented Lagrangian method. Experiments conducted on the weather sentiment dataset and the Duchenne smile dataset indicate that crowdsensing systems become increasingly vulnerable to poisoning as the privacy budget decreases. For example, on the Duchenne smile dataset, lowering the privacy budget from 3.0 to 1.0 resulted in an 86.4% increase in attack effectiveness as measured by Average Utility Damage.

We also propose optimal countermeasures to defend against the proposed data poisoning attacks. The goal is to identify and isolate corrupted workers while fully adhering to LDP protocols. *Since malicious alterations are indistinguishable from the inherent LDP noise perturbation, detecting corrupted workers without breaking LDP protocol is exceptionally challenging.* To overcome this issue, we formulate the defense strategy as an optimization problem aimed at minimizing the deviation introduced by the poisoning attacks. In practice, this means that workers exerting the greatest influence on the deviation are flagged as malicious. We employ the Differential Evolution (DE) algorithm to solve the minimization problem and determine the optimal defense strategy. Evaluations on two real-world datasets confirm the effectiveness of our countermeasures against the proposed poisoning attacks. For instance, on the Duchenne smile dataset, the defense effectiveness (measured by AUC, a standard classifier evaluation metric) drops by only 23.2% when the privacy budget is reduced from 2.8 to 1.0. In summary, our main contributions are as follows:

- We explore the hidden risks posed by LDP in crowdsensing by proposing optimization-based data poisoning attacks and corresponding defenses to LDP-based privacy-preserving crowdsensing systems.
- We model the data poisoning attack as a bi-level optimization problem. Given its NP-hard nature, we decompose it into upper-level and lower-level sub-problems, which are solved iteratively using the augmented Lagrangian method. Our extensive experiments on real-world datasets reveal that LDP perturbation facilitates data poisoning attacks.
- We formulate countermeasures as a minimization problem and solve it using a Differential Evolution algorithm. Our experimental results demonstrate that the proposed defenses significantly mitigate the effects of data poisoning attacks, even when attackers exploit LDP noise perturbation.

The remainder of this chapter is structured as follows. Section 5.2 introduces the problem under investigation and explains two fundamental concepts: truth discovery methods and local differential privacy. Our primary contributions are detailed in Sects. 5.3 and 5.4. In Sect. 5.3, we describe the optimal data poisoning attacks targeting LDP-based privacy-preserving crowdsensing, while Sect. 5.4 presents the corresponding countermeasures. Section 5.5 evaluates the computational complexity, limitations, and performance of our proposed approaches. Finally, Sect. 5.7 concludes this chapter.

5.2 Problem Statement and Basic Concepts

5.2.1 Problem Statement

Attackers capable of manipulating the sensory data sent to the server can compromise the reliability of truth discovery methods, a tactic known as a data poisoning attack. Since the aggregated truth is highly sensitive to even minor alterations in the collected data, attackers can exploit this vulnerability to distort the crowdsensing outcomes. For example, by submitting data that deviates only slightly from the ground truth, attackers can significantly undermine the trustworthiness of the crowdsensing outcomes.

System Model In this chapter, we consider a crowdsensing framework that uses LDP protocols to protect data privacy. The system comprises a server, requesters, and multiple privacy-conscious workers, as shown in Fig. 5.1. The server is assumed to operate honestly while remaining curious about the workers' sensing data, meaning it follows protocols like truth discovery but might misuse private information for profit. Specifically, requesters send sensing tasks, including some objects such as healthcare or travel records, to the crowdsensing server. The server then assigns these tasks to a group of participating workers who perform the sensing and submit perturbed data instead of the actual sensing data to protect their privacy. After gathering all perturbed data, the server uses a truth discovery method to estimate the truths. Finally, these estimated truths are returned to the requesters.

Assume that attackers aim to undermine the crowdsensing system by distorting the truths output by truth discovery methods. As shown in Fig. 5.1, these attackers cannot alter the data provided by honest workers. Instead, they introduce corrupted workers, either fabricated or recruited, to launch attacks by carefully designing their sensory data. Meanwhile, defenders are deployed within the honest-but-curious server to detect such corrupted workers by analyzing the collected data. In this model, attackers and defenders do not communicate. The defenders are unaware of the attackers' tactics, and the attackers remain uninformed about the defensive measures. In this chapter, we investigate data poisoning attacks and their corresponding countermeasures in crowdsensing systems that use LDP to protect workers' sensory data privacy. Note that although worker weights might inadvertently reveal sensitive details such as sensing capabilities or sensor quality, this issue is outside the scope of our discussion.

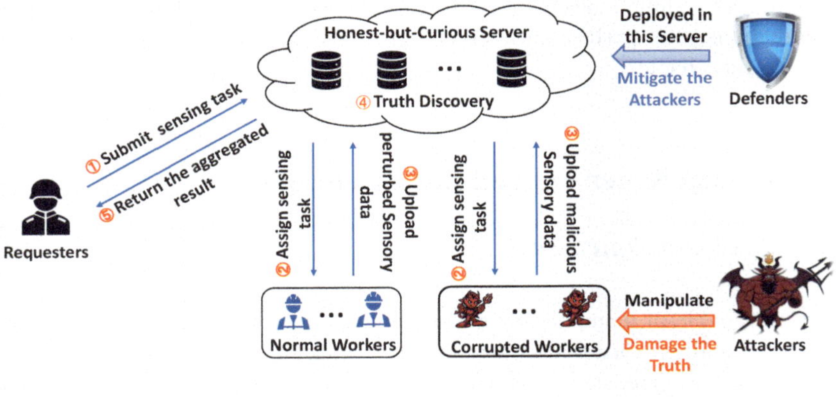

Fig. 5.1 Data poisoning attacks and defenses in LDP-based privacy-preserving crowdsensing systems

We assume that each task comprises $|O|$ objects and is handled by $|U|$ workers. The objects are represented by $O = \{o_1, o_2, \cdots, o_{|O|}\}$ and the workers by $U = \{u_1, u_2, \cdots, u_{|U|}\}$. For any $k \in \mathbb{N}$, let $x_n^m \in \{1, \ldots, k\}$ denote the real sensory data of object o_n provided by worker u_m, and y_n^m represent its perturbed version. The overall collection of sensory data is denoted by $X = \{x_n^m \mid o_n \in O, u_m \in U\}$. For brevity, we define $[k] = \{1, \ldots, k\}$. Worker reliability is captured by weights $W = \{w_1, w_2, \cdots, w_{|U|}\}$, where w_m represents the reliability of worker u_m. As illustrated in Fig. 5.1, the set U is divided into normal workers $\widehat{U} = \{\widehat{u}_1, \cdots, \widehat{u}_{|\widehat{U}|}\}$ and corrupted workers $\widetilde{U} = \{\widetilde{u}_1, \cdots, \widetilde{u}_{|\widetilde{U}|}\}$, so that $U = \widehat{U} \cup \widetilde{U}$. Correspondingly, the weights split into $\widehat{W} = \{\widehat{w}_1, \cdots, \widehat{w}_{|\widehat{U}|}\}$ for normal workers and $\widetilde{W} = \{\widetilde{w}_1, \cdots, \widetilde{w}_{|\widetilde{U}|}\}$ for corrupted ones. The perturbed sensory data provided by privacy-conscious normal workers is denoted by $\widehat{Y} = \{\widehat{y}_n^m \mid o_n \in O, \widehat{u}_m \in \widehat{U}\}$, where each real sensing data $\widehat{x}_n^m$ is perturbed to $\widehat{y}_n^m$. In contrast, corrupted workers directly upload their manipulated data, represented as $\widetilde{X} = \{\widetilde{x}_n^m \mid o_n \in O, \widetilde{u}_m \in \widetilde{U}\}$, without any perturbation.

Attackers' Capability and Background Knowledge Figure 5.1 illustrates that attackers can control a set of corrupted workers to initiate data poisoning attacks by instructing them to submit falsified sensory data. This possibility is realistic given the open nature of crowdsensing systems. As noted in [27], attackers can easily acquire numerous fake accounts on services like Twitter, Google, and Hotmail.

In our model, the attackers are assumed to possess complete knowledge of both the truth discovery method and the LDP protocol used in the crowdsensing system. In addition, the attackers may obtain the real sensory data of some normal workers by methods such as stealing credentials, compromising their systems, or even bribery. Next, we elaborate on two key aspects of the attackers' capabilities and background knowledge.

Definition 5.1 (Percentage of Corrupted Workers) Let $U = \{u_1, u_2, \cdots, u_{|U|}\}$ denote the set of all participating workers, and $\widetilde{U} = \{\widetilde{u}_1, \widetilde{u}_2, \cdots, \widetilde{u}_{|\widetilde{U}|}\}$ represent the subset manipulated by attackers $\widetilde{U} \subset U$. The percentage of corrupted workers is defined as

$$\varphi = |\widetilde{U}|/|U|, \tag{5.1}$$

where $|\widetilde{U}|$ and $|U|$ are the number of corrupted and total workers, respectively.

Definition 5.1 implies that a higher percentage of corrupted workers, represented by φ, equates to increased attacker strength. We focus on the scenario where $0 < \varphi \ll 0.5$ because, when $\varphi > 0.5$ (i.e., corrupted workers outnumber normal ones), data poisoning attacks are essentially guaranteed to succeed.

Definition 5.2 (Percentage of Known Workers) Let attackers have access to the perturbed sensory data of a subset U^{known} of normal workers, where the set of all normal workers is denoted by $\widehat{U}$ and $U^{known} \subset \widehat{U}$. The percentage of known normal workers is then defined as

$$\psi = |U^{known}|/|\widehat{U}|, \tag{5.2}$$

where $|U^{known}|$ is the number of workers known to the attackers and $|\widehat{U}|$ represents the total number of normal workers.

Definition 5.2 indicates that an increased value of ψ (with $0 < \psi \leq 1$), which represents a higher fraction of normal workers whose data is known to the attackers, corresponds to more extensive background knowledge on the part of the attackers.

Attackers' Goals The attackers seek to maximally undermine the utility (or reliability) of the inferred truths while remaining undetected. This strategy is highly pertinent in practical crowdsensing environments. For example, attackers might alter the traffic information on Waze to steer drivers along routes of their choosing [28]. Accordingly, the attackers' objective can be modeled as the following maximization problem:

$$\max_{\widetilde{X}} \ \mathcal{U}\left(\mathcal{A}(\widehat{X}), \widetilde{X}, \mathcal{V}\right), \tag{5.3}$$

where $\mathcal{U}(*)$ quantifies the utility loss (or equivalently, the gain from the attack) regarding the inferred truth, $\mathcal{A}$ represents the LDP protocol, $\widehat{X}$ denotes the real sensory data from normal workers, $\widetilde{X}$ stands for the maliciously altered data provided by corrupted workers, and $\mathcal{V}$ is the truth discovery method.

Defenders' Goals Conversely, defenders aim to develop an optimal countermeasure against these data poisoning attacks as formulated above. Their task is to design a defense mechanism $\mathcal{T}$ that satisfies two main requirements: (1) it preserves the guarantees of LDP, and (2) it minimizes the maximum payoff that attackers can achieve under the privacy constraints. Hence, when the attack strategy $\widetilde{X}$ is predetermined, the problem of finding the optimal defense can be formulated as a minimization problem:

$$\min_{\mathcal{T}} \ \mathcal{U}\left(\mathcal{A}(\widehat{X}), \widetilde{X}, \mathcal{V}, \mathcal{T}\right). \tag{5.4}$$

5.2.2 Basic Concepts

In this section, we introduce fundamental concepts related to truth discovery methods and local differential privacy.

Truth Discovery Methods Truth discovery methods have been widely utilized in crowdsensing to extract reliable insights from the conflicting and noisy sensory data provided by participants [5, 18, 19]. These methods automatically assess each worker's credibility by assigning a corresponding weight and then use these weights to deduce the most trustworthy information from the collected data. The underlying assumptions are twofold: workers consistently providing accurate data receive higher weights, and consequently, the data from these workers is deemed more reliable. In practice, truth discovery is implemented via an iterative algorithm that alternates between two primary steps: truth inference and weight estimation (as outlined in Algorithm 8). The details of these steps are as follows:

- *Truth Inference.* In this phase, the aggregated truth for each object $\forall o_n \in O$ is computed using the current worker weights $W = \{w_1, w_2, \cdots, w_{|U|}\}$, where each worker is denoted by u_m for $m \in \{1, \ldots, |U|\}$. The truth x_n^{trust} for object o_n is obtained through a weighted average:

$$x_n^{trust} = \frac{\sum_{m=1}^{|U|} w_m \cdot x_n^m}{\sum_{m=1}^{|U|} w_m}, \tag{5.5}$$

  where x_n^{truth} is the sensory data provided by worker u_m for object o_n. The resulting value, which is continuous, is then rounded to the nearest integer to establish the final truth (e.g., if $x_n^{truth} = 4.3$, the truth is taken as 4).
- *Weight Estimation.* In the subsequent step, the weights W are updated based on the current set of truth values $X^{truth} = \{x_1^{truth}, x_2^{truth}, \cdots, x_{|O|}^{truth}\}$. A worker's weight is increased if their submitted data closely matches the aggregated truth. This update is typically formulated as:

$$w_m = h\left(\sum_{n=1}^{|O|} d(x_n^m, x_n^{truth})\right), \tag{5.6}$$

  where $h(*)$ is a monotonically decreasing function and $d(x_n^m, x_n^{truth})$ measures the discrepancy between the worker's reported sensing data and the inferred truth.

Algorithm 8 Truth discovery method

Input: Sensory data of $|O|$ objects submitted by U workers $\{x_n^m \mid \forall o_n \in O, \forall u_m \in U\}$; Threshold for convergence $\phi > 0$

Output: Inferred truths $X^{truth} = \{x_1^{truth}, \cdots, x_{|O|}^{truth}\}$ for objects O

```
Initialize worker weights W = {w_m | ∀u_m ∈ U};
X^truth ← Calculate the truths using current weights W based on Eq. (5.5);
while True do
    W ← Calculate the weights using current truths X^truth based on Eq. (5.6);
    X^truth_new = {x^truth_{1,new}, ..., x^truth_{|O|,new}} ← Calculate the truths using current weights W based on Eq. (5.5);
    if 1/|O| · Σ_{n=1}^{|O|} d(x^truth_n, x^truth_{n,new}) ≤ φ then
        break;
    end if
    X^truth ← X^truth_new;
end while
```

We employ the advanced truth discovery method, CRH [5, 19], as an example to demonstrate the core ideas behind optimal data poisoning attacks and the corresponding defenses in LDP-based privacy-preserving crowdsensing. CRH is a well-established optimization-based method, defined by the following formulation:

$$\min_{X^{truth}, W} \sum_{m=1}^{|U|} w_m \cdot \sum_{n=1}^{|O|} d(x_n^m, x_n^{truth}), \tag{5.7a}$$

$$\text{s.t.} \sum_{m=1}^{|U|} \exp(-w_m) = 1, \; |U| > 1. \tag{5.7b}$$

Based on the definition of the worker-weight-updating-function $h(*)$ in CRH, the update rule is given by:

$$w_m = \log\left(\frac{\sum_{m=1}^{|U|} \sum_{n=1}^{|O|} d(x_n^m, x_n^{truth})}{\sum_{n=1}^{|O|} d(x_n^m, x_n^{truth})}\right), \tag{5.8}$$

with the distance function defined as $d(x_n^m, x_n^{truth}) = (x_n^m - x_n^{truth})^2$.

Local Differential Privacy Local Differential Privacy (LDP) [20–22] is a privacy notion designed to protect sensory data privacy in local settings. Essentially, a protocol is said to satisfy LDP if it perturbs any two sensory inputs to the same output with nearly identical probabilities. The formal definition of LDP is presented below.

Definition 5.3 (Local Differential Privacy) Let $k \in \mathbb{N}$ and $[k] = \{1, 2, \cdots, k\}$. A mechanism $\mathcal{A}$ is said to satisfy ε-local differential privacy (ε-LDP) if, for any two sensory inputs $x_{n,1}^m, x_{n,2}^m \in [k]$ and any possible output $y_n^m \in [k]$, the following inequality holds,

$$Pr\left\{\mathcal{A}(x_{n,1}^m) = y_n^m\right\} \leq e^{\varepsilon} \times Pr\left\{\mathcal{A}(x_{n,2}^m) = y_n^m\right\}. \tag{5.9}$$

In this context, $\varepsilon > 0$ represents the privacy budget, and $\mathcal{A}(x_n^m)$ denotes the randomized output of the sensory data x_n^m.

To ensure ε-LDP, a commonly employed technique is the randomized response protocol, specifically Warner's Randomized Response (Warner's RR) [29].

Definition 5.4 (Warner's Randomized Response) Let $k \in \mathbb{N}$ and $[k] = \{1, \cdots, k\}$. For a given sensory input $x_n^m \in [k]$, Warner's Randomized Response produces a perturbed output $y_n^m \in [k]$. Specifically, the mechanism outputs the original value x_n^m with probability p and any other value from $[k]\backslash\{x_n^m\}$ with equal probability q. Thus, the randomized response mechanism is defined as

$$Pr(y_n^m = a) = \begin{cases} \dfrac{e^{\varepsilon}}{k-1+e^{\varepsilon}} := p, & \text{if } a = x_n^m, \\ \dfrac{1}{k-1+e^{\varepsilon}} := q, & \text{otherwise,} \end{cases} \tag{5.10}$$

where ε is the privacy budget.

Directly inferring truth from the perturbed sensory data inevitably introduces bias. Consequently, crowdsensing systems must correct this bias using an unbiased estimate. By applying Warner's Randomized Response (i.e., Eq. (5.10)), we can derive the expected value of the perturbed data y_n^m (which is a random variable) as follows,

$$E[y_n^m] = (p-q) \cdot x_n^m + \frac{(1+k)k}{2} \cdot q. \tag{5.11}$$

Then, we deduce that $x_n^m = \frac{2y_n^m - (1+k)k \cdot q}{2(p-q)}$ since $E[y_n^m] = y_n^m$ (the random variable y_n^m is unbiased). Therefore, the unbiased estimate of truth x_n^{truth} is updated as follows

$$x_n^{truth} = \frac{-(1+k)k \cdot q + \sum_{m=1}^{|U|} 2w_m \cdot y_n^m}{2(p-q) \cdot \sum_{m=1}^{|U|} w_m}. \tag{5.12}$$

This equation reveals that the calibrated truth x_n^{truth} is highly sensitive to even minor changes in the perturbed sensory data. The reason is that $p - q$ is a value less than 1, i.e., $0 < p - q < 1$, which effectively amplifies the weight of each worker, i.e., $w_m/(p-q) > w_m$.

5.3 Data Poisoning Attacks Hidden Behind the LDP Noise

The calibration process in LDP protocols, which adjusts for bias introduced by LDP noise, can inadvertently amplify the impact of malicious sensory data on inferred truths. As a result, malicious data with slight deviations from normal sensory data can manipulate the inferred truths while evading detection by truth discovery methods. Building on this insight, we introduce a novel attack strategy called "hiding behind the LDP noise" and develop an optimization-based data poisoning attack. Specifically, we analyze LDP-based privacy-preserving truth discovery methods, which leverage the calibration to correct the bias induced by LDP noise perturbation, deriving truths from the collected dataset that includes both malicious and LDP-perturbed data (detailed in Sect. 5.3.1). To exploit this, attackers formulate the data poisoning attack as a bi-level optimization problem, aiming to maximize deviation in inferred truths given the constraints of privacy-preserving truth discovery methods (detailed in Sect. 5.3.2). This allows attackers to achieve significant attack gains while maintaining stealth. Finally, we employ an optimal poisoning attack algorithm that combines alternating optimization with the augmented Lagrangian method to solve the formulated problem, yielding the optimal attack strategies (detailed in Sect. 5.3.3).

5.3.1 LDP-Based Privacy-Preserving Truth Discovery Methods

Figure 5.1 illustrates that the crowdsensing server receives two types of sensory data: perturbed data from normal workers and malicious data from compromised ones. To extract the true information from this mix, we reformulate the CRH framework (detailed in Sect. 5.2.2) as follows:

$$P_l: \quad \min_{X_a^{truth}, \widehat{W}, \widetilde{W}} \quad \sum_{m=1}^{|\widehat{U}|} \widehat{w}_m \cdot \sum_{n=1}^{|O|} d\left(\widehat{y}_n^m, x_{a,n}^{truth}\right) + \sum_{m=1}^{|\widetilde{U}|} \widetilde{w}_m \cdot \sum_{n=1}^{|O|} d\left(\widetilde{x}_n^m, x_{a,n}^{truth}\right), \tag{5.13a}$$

$$\text{s.t.} \quad \sum_{m=1}^{|\widehat{U}|} \exp(-\widehat{w}_m) + \sum_{m=1}^{|\widetilde{U}|} \exp(-\widetilde{w}_m) = 1, \tag{5.13b}$$

where $X_a^{truth} = \{x_{a,1}^{truth}, \cdots, x_{a,|O|}^{truth}\}$ denotes the inferred truths after the attack, and problem P_l is solved using Algorithm 8.

Next, we update both the truth and worker-weight functions (i.e., Eqs. (5.5) and (5.6) in Algorithm 8). To comply with ε-LDP, each normal worker implements Warner's Randomized Response (RR) locally (detailed in Definition 5.4), i.e.,

$$Pr(\widehat{y}_n^m = a) = \begin{cases} \dfrac{e^{\varepsilon}}{k-1+e^{\varepsilon}} := p, & \text{if } a = \widehat{x}_n^m, \\ \dfrac{1}{k-1+e^{\varepsilon}} := q, & \text{otherwise.} \end{cases} \tag{5.14}$$

According to the Eq. (5.14), the expectation is computed as $E[\widehat{y}_n^m] = (p-q)\cdot\widehat{x}_n^m + \frac{(1+k)k}{2}\cdot q$. Due to $E[\widehat{y}_n^m] = \widehat{y}_n^m$, we have $\widehat{x}_n^m = \frac{2\widehat{y}_n^m - (1+k)k\cdot q}{2(p-q)}$. Thus, the unbiased estimation of $x_{a,n}^{truth}$ is updated as

$$x_{a,n}^{truth} = T(\widehat{W}, \widetilde{W}, \widehat{Y}, \widetilde{X}) \tag{5.15}$$

$$= \frac{-(1+d)k\cdot q + \sum_{m=1}^{|\widehat{U}|} 2\widehat{w}_m\widehat{y}_n^m + \sum_{m=1}^{|\widetilde{U}|} 2\widetilde{w}_m\widetilde{x}_n^m}{2(p-q)\cdot\left(\sum_{m=1}^{|\widehat{U}|}\widehat{w}_m + \sum_{m=1}^{|\widetilde{U}|}\widetilde{w}_m\right)}. \tag{5.16}$$

This new updating rule reveals that $x_{a,n}^{truth}$ is predominantly influenced by data from higher-weight workers and is sensitive to minor alterations in the malicious inputs $\widetilde{x}_n^m$.

Similarly, we adjust the worker-weight updating mechanism. The weights for normal workers ($\forall \widehat{w}_m \in \widehat{W}$) and corrupted workers $\forall \widetilde{w}_m \in \widetilde{W}$ are updated via

$$\widehat{w}_m = \log \frac{\sum_{n=1}^{|O|}(\sum_{m=1}^{|\widehat{U}|} d(\widehat{y}_n^m, x_{a,n}^{truth}) + \sum_{m=1}^{|\widetilde{U}|} d(\widetilde{x}_n^m, x_{a,n}^{truth}))}{\sum_{n=1}^{|O|} d(\widehat{x}_n^m, x_{a,n}^{truth})}, \tag{5.17}$$

$$\widetilde{w}_m = \log \frac{\sum_{n=1}^{|O|}(\sum_{m=1}^{|\widehat{U}|} d(\widehat{y}_n^m, x_{a,n}^{truth}) + \sum_{m=1}^{|\widetilde{U}|} d(\widetilde{x}_n^m, x_{a,n}^{truth}))}{\sum_{n=1}^{|O|} d(\widetilde{x}_n^m, x_{a,n}^{truth})}. \tag{5.18}$$

According to Eq. (5.18), corrupted workers receive higher weights if their submitted data is closer to the truth. That is, a smaller value of $d(\widetilde{x}_n^m, x_{a,n}^{truth})$ results in a larger weight for the corrupted worker $\widetilde{u}_m$.

By integrating these revised updating rules (i.e., Eqs. (5.16)–(5.18)), we arrive at the updated Algorithm 8, which ultimately allows us to infer the post-attack truths X_a^{truth} (namely the truths after attacks).

5.3.2 Formulating Optimal Data Poisoning Attacks

Section 5.2.1 explains that attackers seek to design optimal data poisoning attacks that both (1) maximize the deviation of the inferred truths and (2) remain undetected. Here, the attack damage is measured by the total alteration in the inferred truths. To maximize this damage, attackers target the cumulative alteration $\sum_{n=1}^{|O|} sgn(x_{a,n}^{truth} - x_n^{truth})$ when they intend $x_{a,n}^{truth}$ to far exceed x_n^{truth} (i.e., $x_{a,n}^{truth} \gg x_n^{truth}$), or $\sum_{n=1}^{|O|} sgn(x_n^{truth} - x_{a,n}^{truth})$ when they aim for $x_{a,n}^{truth}$ to be substantially lower than x_n^{truth} (i.e., $x_{a,n}^{truth} \ll x_n^{truth}$). The signum function $sgn(x_{a,n}^{truth} - x_n^{truth})$ is formulated as

$$sgn(x_{a,n}^{truth} - x_n^{truth}) = \begin{cases} 1, & \text{if } x_{a,n}^{truth} > x_n^{truth}, \\ 0, & \text{if } x_{a,n}^{truth} = x_n^{truth}, \\ -1, & \text{if } x_{a,n}^{truth} < x_n^{truth}. \end{cases} \tag{5.19}$$

The truths denoted by $X^{truth} = \{x_1^{truth}, x_2^{truth}, \cdots, x_{|O|}^{truth}\}$ are derived by processing the sensory data provided by the known workers U^{known} using Algorithm 8 (detailed in Sect. 5.2.2). Consequently, the attackers' payoff is defined as either $\mathcal{U}(\mathcal{A}(\widehat{X}), \widetilde{X}, \mathcal{V}) = \sum_{n=1}^{|O|} sgn(x_{a,n}^{truth} - x_n^{truth})$ or $\mathcal{U}(\mathcal{A}(\widehat{X}), \widetilde{X}, \mathcal{V}) =$

$\sum_{n=1}^{|O|} sgn(x_n^{truth} - x_{a,n}^{truth})$ depending on whether the goal is to significantly raise or lower the inferred truths.

Corrupted workers that are identified by truth discovery methods (e.g., CRH) receive reduced weights, thereby diminishing their influence on the overall aggregation. Therefore, evading detection by these methods becomes essential for attackers. To achieve this, we incorporate the optimization problem P_l (refer to Eq. (5.13a)) as a constraint. Additionally, to limit the search space for the attack strategy $\widetilde{X}$, we confine the malicious sensory data to the range of genuine sensory data available to the attackers. Formally, for every object $\forall o_n \in O$ and each corrupted worker $\widetilde{u}_m \in \widetilde{U}$, the manipulated sensory data $\widetilde{x}_n^m$ must lie within $[k']$, i.e., $\widetilde{x}_n^m \in [k']$, where k' is the highest sensory value known to attackers. Based on these considerations, we formulate the dual objectives as the following optimization problem,

$$P: \max_{\widetilde{X}=\{\widetilde{x}_n^m | o_n \in O, \widetilde{u}_m \in \widetilde{U}\}} \sum_{n=1}^{|O|} sgn(x_{a,n}^{truth} - x_n^{truth}), \tag{5.20a}$$

$$\text{s.t. } \widetilde{x}_n^m \in [k'], \quad \forall o_n \in O, \ \forall \widetilde{u}_m \in \widetilde{U}, \tag{5.20b}$$

$$\{X_a^{truth}, \widehat{W}, \widetilde{W}\} := \arg\min_{X_a^{truth}, \widehat{W}, \widetilde{W}} \left\{ \sum_{m=1}^{|\widehat{U}|} \widehat{w}_m \sum_{n=1}^{|O|} d\left(\widehat{y}_n^m, x_{a,n}^{truth}\right) + \sum_{m=1}^{|\widetilde{U}|} \widetilde{w}_m \cdot \sum_{n=1}^{|O|} d\left(\widetilde{x}_n^m, x_{a,n}^{truth}\right) \right\}, \tag{5.20c}$$

$$\sum_{m=1}^{|\widehat{U}|} \exp(-\widehat{w}_m) + \sum_{m=1}^{|\widetilde{U}|} \exp(-\widetilde{w}_m) = 1. \tag{5.20d}$$

This formulated problem P (detailed in Eq. (5.20a)) is a bi-level optimization problem [30] because its constraints include the inner optimization problem P_l. In this structure, the upper-level problem optimizes $\widetilde{X}$, while the lower-level problem determines $(X_a^{truth}, \widehat{W}, \widetilde{W})$ given $\widetilde{X}$. Note that the attackers' objective is to increase the inferred truth values. If their aim were to decrease these values instead, the objective function in Eq. (5.20a) would be altered to

$$\max_{\widetilde{X}=\{\widetilde{x}_n^m | o_n \in O, \widetilde{u}_m \in \widetilde{U}\}} \sum_{n=1}^{|O|} sgn(x_n^{truth} - x_{a,n}^{truth}). \tag{5.21}$$

Moreover, the problem P forms a general attack framework applicable to any optimization-based truth discovery approach by modifying the lower-level problem accordingly.

In summary, by formulating data poisoning as a bi-level maximization problem, attackers can determine an effective strategy that maximizes the deviation of inferred truths while conforming to the constraints of LDP-based privacy-preserving truth discovery methods. This allows attackers to distort the final outputs of the truth discovery methods to the greatest extent possible.

5.3.3 Finding Optimal Data Poisoning Attacks

Due to the NP-hard nature of bi-level optimization [30], directly solving problem P is impractical. Therefore, we adopt an alternating optimization approach [9] that iteratively tackles the upper-level and lower-level sub-problems. With the lower-level problem P_l already addressed in Sect. 5.3.1, our focus shifts to the upper-level sub-problem P_u. Given worker weights $W = \widetilde{W} \cup \widehat{W}$, the upper-level problem is formulated as

$$P_u: \max_{\widetilde{X}} \sum_{n=1}^{|O|} sgn(x_{a,n}^{truth} - x_n^{truth}), \tag{5.22a}$$

$$\text{s.t. } \widetilde{x}_n^m \in [k'], \ \forall o_n \in O, \ \forall \widetilde{u}_m \in \widetilde{U}, \tag{5.22b}$$

$$x_{a,n}^{truth} = T(\widehat{W}, \widetilde{W}, \widehat{Y}, \widetilde{X}), \forall o_n \in O, \tag{5.22c}$$

where $T(\cdot)$ is given in Eq. (5.16) and x_n^{truth} is derived from the attackers' background knowledge.

Since the discrete nature of the signum function (i.e., Eq. (5.22a)) and the constraint (i.e., Eq. (5.22b)) makes P_u difficult to solve, as shown in Fig. 5.2, we

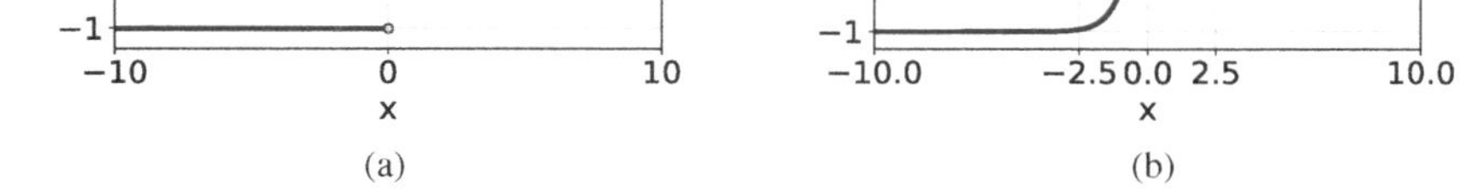

Fig. 5.2 The piecewise function $sgn(x)$ is relaxed into the continuous function $\tanh(x)$. (**a**) $sgn(x)$. (**b**) $\tanh(x)$

first relax the signum function (as defined in Eq. (5.19)) into the continuous function $\tanh(*)$, i.e.,

$$\tanh(x_{a,n}^{truth} - x_n^{truth}) = \frac{e^{x_{a,n}^{truth} - x_n^{truth}} - e^{x_n^{truth} - x_{a,n}^{truth}}}{e^{x_{a,n}^{truth} - x_n^{truth}} + e^{x_n^{truth} - x_{a,n}^{truth}}}. \tag{5.23}$$

Then, we relax the constraint Eq. (5.22b) to $1 \le \widetilde{x}_n^m \le k', \forall o_n \in O, \forall \widetilde{u}_m \in \widetilde{U}$. By this way, we yield the continuous problem P_u^1, i.e.,

$$P_u^1 : \max_{\widetilde{X}} \sum_{n=1}^{|O|} \tanh(x_{a,n}^{truth} - x_n^{truth}), \tag{5.24}$$

$$\text{s.t. } 1 \le \widetilde{x}_n^m \le k', \forall o_n \in O, \forall \widetilde{u}_m \in \widetilde{U}, \tag{5.25}$$

$$x_{a,n}^{truth} = T(\widehat{W}, \widetilde{W}, \widehat{Y}, \widetilde{X}), \forall o_n \in O. \tag{5.26}$$

The problem P_u is designed to reverse the inferred truth x_n^{truth} for every object $\forall o_n \in O$, whereas P_u^1 not only reverses but also pushes the inferred value further away from x_n^{truth}. In other words, replacing the original objective (Eq. (5.22a)) with the one in Eq. (5.24) effectively intensifies the attack goal.

Besides, since the decision variable $\widetilde{X} = \{\widetilde{x}_n^m \mid \forall o_n \in O, \forall \widetilde{u}_m \in \widetilde{U}\}$ is an $|O| \times |\widetilde{U}|$ dimensional matrix, the computational complexity of solving problem P_u^1 increases sharply with the number of corrupted workers $|\widetilde{U}|$. To address this, we assume that all corrupted workers report the same malicious data, i.e., for any two corrupted workers, $\widetilde{X}^i = \widetilde{X}^j$, where $\widetilde{X}^i = \{\widetilde{x}_1^i, \cdots, \widetilde{x}_{|O|}^i\}$ represents the data from worker $\widetilde{u}_i$. This assumption is practical, as attackers maximize their success when every corrupted worker exhibits identical behavior. Moreover, since P_u^1 maximizes $\tanh(x_{a,n}^{truth} - x_n^{truth})$, it guarantees that $x_{a,n}^{truth} \ge x_n^{truth}$ for all $o_n \in O$. Accordingly, we can set $\widetilde{X} = [\widetilde{X}^i, \cdots, \widetilde{X}^i]_{|\widetilde{U}|}$, thereby relaxing the original problem P_u to the following problem P_u^2, i.e.,

$$P_u^2 : \max_{\widetilde{X}^i = \{\widetilde{x}_1^i, \cdots, \widetilde{x}_{|O|}^i\}} \sum_{n=1}^{|O|} \tanh(x_{a,n}^{truth} - x_n^{truth}), \tag{5.27a}$$

$$\text{s.t. } 1 \le \widetilde{x}_n^i \le k', \forall o_n \in O. \tag{5.27b}$$

$$x_{a,n}^{truth} = T(\widehat{W}, \widetilde{W}, \widehat{Y}, \widetilde{X}), \forall o_n \in O, \tag{5.27c}$$

$$x_{a,n}^{truth} \ge x_n^{truth}, \forall o_n \in O. \tag{5.27d}$$

In what follows, we prove that the problem P_u^2 is a convex optimization problem.

Theorem 5.1 *The problem P_u^2 is a convex optimization problem.*

Proof The problem P_u^2 is convex if and only if the following minimization formulation is convex,

$$\min_{\widetilde{X}^i} \quad -\sum_{n=1}^{|O|} \tanh(x_{a,n}^{truth} - x_n^{truth}), \tag{5.28a}$$

$$\text{s.t.} \quad \widetilde{x}_n^i \leq k', -\widetilde{x}_n^i \leq -1, \forall o_n \in O, \tag{5.28b}$$

$$x_{a,n}^{truth} = T(\widehat{W}, \widetilde{W}, \widehat{Y}, \widetilde{X}), \forall o_n \in O, \tag{5.28c}$$

$$-x_{a,n}^{truth} \leq -x_n^{truth}, \forall o_n \in O. \tag{5.28d}$$

Since both the decision variable $\widetilde{x}_n^i$ and the function $T(\widehat{W}, \widetilde{W}, \widehat{Y}, \widetilde{X})$ (defined in Eq. (5.16)) are linear with respect to $\widetilde{x}_n^i$ for every $o_n \in O$, the feasible region defined by constraints (5.28b)–(5.28d) is convex. Next, we examine the convexity of the objective function. Let $g(x) = -\tanh(x)$. Its first derivative is $g'(x) = g^2(x) - 1 \leq 0$, and its second derivative is $g''(x) = 2g(x) \cdot g'(x)$. Because $g(x) \leq 0$ when $x \geq 0$, it follows that $g''(x) \geq 0$ for $x \geq 0$, i.e., $g(x)$ is a monotonically decreasing convex function on the interval $[0, +\infty)$. Moreover, considering the composition $g(x_{a,n}^{truth} - x_n^{truth})$ and applying the chain rule, we have

$$\frac{\partial^2 g(x_{a,n}^{truth} - x_n^{truth})}{\partial(\widetilde{x}_n^i)^2}$$
$$= g''(x_{a,n}^{truth} - x_n^{truth}) \cdot \left((|\widetilde{U}| \cdot \widetilde{w}_i)/((p-q) \cdot (\sum_{m=1}^{|\widehat{U}|} \widehat{w}_m + |\widehat{U}| \cdot \widetilde{w}_i)) \right)^2. \tag{5.29}$$

Since $g''(x) \geq 0$ for $x \geq 0$ and the derivative of the inner function is constant, it follows that the second derivative with respect to $\widetilde{x}_n^i$ is nonnegative whenever $x_{a,n}^{truth} \geq x_n^{truth}$. Thus, the function $\sum_{n=1}^{|O|} g(x_{a,n}^{truth} - x_n^{truth})$ is convex in $\widetilde{X}^i$ under the constraint $x_{a,n}^{truth} \geq x_n^{truth}, \forall o_n \in O$ (which is enforced by (5.28d)). In summary, the convexity of both the feasible set and the objective function confirms that the minimization problem is convex, and therefore, problem P_u^2 is a convex optimization problem. □

Since problem P_u^2 is convex, we can solve it using a gradient-based augmented Lagrangian approach. To transform P_u^2 into an equivalent formulation, we introduce slack variables $S = \{s_1, s_2, \cdots, s_{|O|}\}$, $S' = \{s'_1, s'_2, \cdots, s'_{|O|}\}$, and $S'' = \{s''_1, s''_2, \cdots, s''_{|O|}\}$, which leads to the equivalent problem, i.e.,

$$\min_{\widetilde{X}^i} \sum_{n=1}^{|O|} -\tanh(x_{a,n}^{truth} - x_n^{truth}), \tag{5.30a}$$

$$\text{s.t.} \quad 1 - \widetilde{x}_n^i + s_n = 0, \ \forall o_n \in O, \tag{5.30b}$$

$$\widetilde{x}_n^i - k' + s'_n = 0, \ \forall o_n \in O, \tag{5.30c}$$

$$x_{a,n}^{truth} - x_n^{truth} + s_n'' = 0, \forall o_n \in O, \tag{5.30d}$$

$$x_{a,n}^{truth} - T(\widehat{W}, \widetilde{W}, \widehat{Y}, \widetilde{X}) = 0, \forall o_n \in O, \tag{5.30e}$$

$$s_n, s_n', s_n'' \geq 0, \ \forall o_n \in O. \tag{5.30f}$$

We then construct the augmented Lagrangian function of above optimization problem Eq. (5.30a), i.e.,

$$\begin{aligned} &L_\sigma(\widetilde{X}^i, S, S', S'', \lambda, \mu, \nu) \\ &= \sum_{n=1}^{|O|} -\tanh(x_{a,n}^{truth} - x_n^{truth}) + \frac{\sigma}{2} \cdot \rho(\widetilde{X}^i, S, S', S'') \\ &\quad + \sum_{n=1}^{|O|} \lambda_n \cdot (1 - \widetilde{x}_n^i + s_n) + \sum_{n=1}^{|O|} \mu_n \cdot (\widetilde{x}_n^i - k' + s_n') \\ &\quad + \sum_{n=1}^{|O|} \nu_n \cdot (x_{a,n}^{truth} - x_n^{truth} + s_n''), \end{aligned} \tag{5.31}$$

where $\lambda = \{\lambda_1, \cdots, \lambda_{|O|}\}$, $\mu = \{\mu_1, \cdots, \mu_{|O|}\}$, and $\nu = \{\nu_1, \cdots, \nu_{|O|}\}$ are Lagrange multipliers, σ is the penalty factor, and the quadratic penalty function $\rho(\widetilde{X}^i, S, S', S'')$ is formulated as

$$\begin{aligned} &\rho(\widetilde{X}^i, S, S', S'') \\ &= \sum_{n=1}^{|O|} \left(1 - \widetilde{x}_n^i + s_n\right)^2 + \sum_{n=1}^{|O|} \left(\widetilde{x}_n^i - k' + s_n'\right)^2 + \sum_{n=1}^{|O|} (x_{a,n}^{truth} - x_n^{truth} + s_n'')^2. \end{aligned} \tag{5.32}$$

By fixing $\widetilde{X}^i$, λ, μ, and ν, we can decouple the minimization over the slack variables. Their sub-problems are

$$\min_{S \geq 0} \sum_{n=1}^{|O|} \lambda_n \cdot (1 - \widetilde{x}_n^i + s_n) + \frac{\sigma}{2} \cdot \sum_{n=1}^{|O|} (1 - \widetilde{x}_n^i + s_n)^2, \tag{5.33}$$

$$\min_{S' \geq 0} \sum_{n=1}^{|O|} \mu_n \cdot (\widetilde{x}_n^i - k' + s_n') + \frac{\sigma}{2} \cdot \sum_{n=1}^{|O|} (\widetilde{x}_n^i - k' + s_n')^2, \tag{5.34}$$

$$\min_{S'' \geq 0} \sum_{n=1}^{|O|} \nu_n \cdot (x_{a,n}^{truth} - x_n^{truth} + s_n'') + \frac{\sigma}{2} \cdot \sum_{n=1}^{|O|} (x_{a,n}^{truth} - x_n^{truth} + s_n'')^2. \tag{5.35}$$

Therefore, we have

$$s_n = \max\left\{-\frac{\lambda_n}{\sigma} - 1 + \widetilde{x}_n^i, 0\right\}, \ \forall o_n \in O, \tag{5.36}$$

$$s'_n = \max\left\{-\frac{\mu_n}{\sigma} + k' - \widetilde{x}_n^i, 0\right\}, \ \forall o_n \in O, \tag{5.37}$$

$$s''_n = \max\left\{-\frac{\nu_n}{\sigma} - x_{a,n}^{truth} + x_n^{truth}, 0\right\}, \forall o_n \in O. \tag{5.38}$$

The slack variables S, S', and S'' can be eliminated by bringing the Eqs. (5.36)–(5.38) into the augmented Lagrangian function Eq. (5.31), and then we obtain a new augmented Lagrangian function $L_\sigma(\widetilde{X}^i, \lambda, \mu, \nu)$. Finally, the gradient of the augmented Lagrangian function $L_\sigma(\widetilde{X}^i, \lambda, \mu, \nu)$ with respect to $\widetilde{X}^i$, i.e., $\nabla_{\widetilde{X}^i} L_\sigma(\widetilde{X}^i, \lambda, \mu, \nu) = [\frac{\partial L_{\sigma_k}}{\partial \widetilde{x}_1^i}, \cdots, \frac{\partial L_{\sigma_k}}{\partial \widetilde{x}_n^i}]_{|O|}$, can be computed as follows,

$$\begin{aligned}
\frac{\partial L_\sigma(\widetilde{X}^i, \lambda, \mu, \nu)}{\partial \widetilde{x}_n^i} &= ((\tanh(x_{a,n}^{truth} - x_n^{truth}))^2 - 1) \cdot \frac{\partial x_{a,n}^{truth}}{\partial \widetilde{x}_n^i} \\
&+ \sigma \cdot ((1 - \widetilde{x}_n^i + s_n) \cdot (-1 + \frac{\partial s_n}{\partial \widetilde{x}_n^i}) + (\widetilde{x}_n^i - k' + s'_n) \cdot (1 + \frac{\partial s'_n}{\partial \widetilde{x}_n^i}) \\
&+ (x_{a,n}^{truth} - x_n^{truth} + s''_n) \cdot (\frac{\partial x_{a,n}^{truth}}{\partial \widetilde{x}_n^i} + \frac{\partial s''_n}{\partial \widetilde{x}_n^i})) + \lambda_n \cdot (-1 + \frac{\partial s_n}{\partial \widetilde{x}_n^i}) \\
&+ \mu_n \cdot (1 + \frac{\partial s'_n}{\partial \widetilde{x}_n^i}) + \nu_n \cdot (\frac{x_{a,n}^{truth}}{\partial \widetilde{x}_n^i} + \frac{\partial s''_n}{\partial \widetilde{x}_n^i}), \ \forall o_n \in O,
\end{aligned} \tag{5.39}$$

where

$$\frac{\partial x_{a,n}^{truth}}{\partial \widetilde{x}_n^i} = \frac{\sum_{m=1}^{|\widetilde{U}|} \widetilde{w}_m}{(p-q) \cdot (\sum_{m=1}^{|\widetilde{U}|} \widehat{w}_m + \sum_{m=1}^{|\widehat{U}|} \widetilde{w}_m)}, \tag{5.40}$$

$$\frac{\partial s_n}{\partial \widetilde{x}_n^i} = \begin{cases} 1, & s_n > 0, \\ 0, & s_n = 0, \end{cases} \tag{5.41}$$

$$\frac{\partial s'_n}{\partial \widetilde{x}_n^i} = \begin{cases} -1, & s'_n > 0, \\ 0, & s'_n = 0, \end{cases} \tag{5.42}$$

$$\frac{\partial s''_n}{\partial \widetilde{x}_n^i} = \begin{cases} -\frac{\partial x_{a,n}^{truth}}{\partial \widetilde{x}_n^i}, & s''_n > 0, \\ 0, & s''_n = 0. \end{cases} \tag{5.43}$$

Algorithm 9 Optimal poisoning attack algorithm

Input: Lagrange multipliers λ^0, μ^0, and ν^0; Penalty factor $\sigma_0 > 0$; Threshold for convergence $\eta > 0$; Constraint violation constant $\tau > 0$; Constants $0 < \alpha \leq \beta \leq 1$ and $\rho > 1$; Step size ϑ

Output: Optimal poisoning attack $\widetilde{X} = [\widetilde{X}^i, \cdots, \widetilde{X}^i]_{|\widetilde{U}|}$

Initialize the poisoning attack strategy $\widetilde{x} \sim U(1, k'), \forall \widetilde{x} \in \widetilde{X}^{i,0}$, $\eta_0 = \frac{1}{\sigma_0}$, and $\tau_0 = \frac{1}{\sigma_0^\alpha}$;
for $k = 0, 1, 2, 3, \cdots$ **do**
  while $\|\nabla_{\widetilde{X}^i} L_{\sigma_k}(\widetilde{X}^{i,k}, \lambda^k, \mu^k, \nu^k)\|_2 \leq \eta_k$ **do**
    $\{X_a^{truth}, \widehat{W}^k, \widetilde{W}^k\}$ $\leftarrow$ Solving the lower-level problem P_l using $[\widetilde{X}^{i,k}, \cdots, \widetilde{X}^{i,k}]_{|\widetilde{U}|}$ by Algorithm 8;
    $\nabla_{\widetilde{X}^i} L_{\sigma_k}(\widetilde{X}^{i,k}, \lambda^k, \mu^k, \nu^k) = [\frac{\partial L_\sigma}{\partial \widetilde{x}_1^i}, \cdots, \frac{\partial L_\sigma}{\partial \widetilde{x}_{|O|}^i}]$ $\leftarrow$ Computing the gradient according to Eq. (5.39) using X_a^{truth}, $\widehat{W}^k$, $\widetilde{W}^k$;
    $\widetilde{X}^{i,k+1} \leftarrow \widetilde{X}^{i,k} + \vartheta \cdot \nabla_{\widetilde{X}^i} L_{\sigma_k}(\widetilde{X}^{i,k}, \lambda^k, \mu^k, \nu^k)$;
    $\widetilde{X}^{i,k} \leftarrow \widetilde{X}^{i,k+1}$;
  end while
  if $v_k(\widetilde{X}^{i,k+1}) \leq \tau_k$ **then**
    if $v_k(\widetilde{X}^{i,k+1}) \leq \tau$ and $\|\nabla_{\widetilde{X}^i} L_{\sigma_k}(\widetilde{X}^{i,k+1}, \lambda^k, \mu^k, \nu^k)\|_2 \leq \eta$ **then**
      Return $\widetilde{X}^{i,k+1}, \lambda^k, \mu^k, \nu^k$;
    end if
    Update Lagrange multiplier λ_n^{k+1} based on $\max\{\lambda_n^k + \sigma_k \cdot (1 - \widetilde{x}_n^{i,k+1}), 0\}$, $\forall o_n \in O$;
    Update Lagrange multiplier μ_n^{k+1} based on $\max\{\mu_n^k + \sigma_k \cdot (\widetilde{x}_n^{i,k+1} - k'), 0\}$, $\forall o_n \in O$;
    Update Lagrange multiplier ν_n^{k+1} based on $\max\{\nu_n^k + \sigma_k \cdot (x_{a,n}^{truth} - x_n^{truth}), 0\}$, $\forall o_n \in O$;
    Keep penalty factor unchanged: $\sigma_{k+1} \leftarrow \sigma_k$;
    Reduce the threshold for convergence and the constraint violation constant of sub-problem based on $\eta_{k+1} \leftarrow \frac{\eta_k}{\sigma_{k=1}}$, $\tau_{k+1} \leftarrow \frac{\tau_k}{\sigma_{k+1}^\beta}$;
  else
    Keep Lagrange multipliers unchanged: $\lambda^{k+1} \leftarrow \lambda^k$, $\mu^{k+1} \leftarrow \mu^k$, and $\nu^{k+1} \leftarrow \nu^k$;
    Update penalty factor σ_{k+1} based on $\rho \cdot \sigma_k$;
    Adjust the threshold for convergence and the constraint violation constant of the sub-problem based on $\eta_{k+1} \leftarrow \frac{1}{\sigma_{k+1}}$, $\tau_{k+1} \leftarrow \frac{1}{\sigma_{k+1}^\alpha}$;
  end if
end for

Based on the preceding discussion, we propose an optimal poisoning attack algorithm that integrates alternating optimization with the augmented Lagrangian method (see Algorithm 9) to solve problem P. Initially, the attack strategy is set as $\widetilde{X} = [\widetilde{X}^i, \cdots, \widetilde{X}^i]_{|\widetilde{U}|}$, with each element in $\widetilde{X}^i$ sampled uniformly from

$U(1, k')$. In each iteration, we update the solution $\widetilde{X}^{i,k+1}$ by alternating between the following steps using the current penalty parameter σ_k and Lagrange multipliers $\lambda^k = \{\lambda_1^k, \cdots, \lambda_{|O|}^k\}$, $\mu^k = \{\mu_1^k, \cdots, \mu_{|O|}^k\}$, and $\nu^k = \{\nu_1^k, \cdots, \nu_{|O|}^k\}$:

- Lower-Level Update: With the current attack strategy $\widetilde{X} = [\widetilde{X}^i, \widetilde{X}^i, \cdots, \widetilde{X}^i]_{|\widetilde{U}|}$ fixed, update the inferred truth and worker weights $\{\widetilde{X}_a^{truth}, \widetilde{W}, \widehat{W}\}$ by solving the lower-level sub-problem P_l (line 4);
- Gradient Calculation: With $\{\widetilde{X}_a^{truth}, \widetilde{W}, \widehat{W}\}$ fixed, compute the gradient of the augmented Lagrangian $L_{\sigma_k}(\widetilde{X}^i, \lambda^k, \mu^k, \nu^k)$ with respect to $\widetilde{X}^i$ as given in Eq. (5.39) (line 5);
- Strategy Update: Update the attack strategy using a gradient step

$$\widetilde{X}^i + \vartheta \cdot \nabla_{\widetilde{X}^i} L_{\sigma_k}(\widetilde{X}^i, \lambda^k, \mu^k, \nu^k), \tag{5.44}$$

  where ϑ denotes the step size (line 6).

After these steps, we assess the degree of constraint violation, $\upsilon_k(\widetilde{X}^{i,k+1})$, defined as

$$\begin{aligned} &\upsilon_k(\widetilde{X}^{i,k+1}) \\ &= \left(\sum_{n=1}^{|O|} (1 - x_n^{i,k+1} + s_n)^2 + \sum_{n+1}^{|O|} (x_n^{i,k+1} - k' + s_n')^2 \right. \\ &\left. + \sum_{n=1}^{|O|} (x_{a,n}^{truth} - x_n^{truth} + s_n'')^2 \right)^{\frac{1}{2}}, \end{aligned} \tag{5.45}$$

with s_n, s_n', and s_n'' are calculated according to Eqs. (5.36)–(5.38). If the violation $\upsilon_k(\widetilde{X}^{i,k+1})$ meets the desired accuracy (line 9), the algorithm updates the Lagrange multipliers λ^{k+1}, μ^{k+1}, and ν^{k+1} (lines 13, 14, and 15) and refines the solution accuracy (line 17), while keeping the penalty parameter σ_{k+1} unchanged (line 16). Otherwise, the multipliers remain unchanged (line 19), and the penalty factor σ_{k+1} is increased (line 20) to further reduce the constraint violation (line 21). This alternating procedure, combined with the augmented Lagrangian method, iteratively refines the attack strategy until convergence.

5.4 Countermeasures: Designing Optimal Defenses

5.4.1 Formulating Optimal Countermeasures

As described in Sect. 5.2.1, the defenders' objective is to counteract data poisoning attacks. In contrast, the attackers aim to severely distort the inferred reality by directing some compromised workers to send falsified sensory data. By examining the sensory data received at the server, defenders can pinpoint the workers whose data most adversely affects the inferred results and flag them as compromised. To model this, we introduce a set of corruption probabilities $P_D = \{p_1, p_2, \cdots, p_{|U|}\}$, where each p_m (with $0 \leq p_m \leq 1$) quantifies the likelihood that worker u_m has been compromised. Assuming that the defenders know the total number N of workers that attackers can corrupt, the set of detected compromised workers is defined as

$$U_D = \{u_m | \forall p_m \in sort(P_D, N)\}, \tag{5.46}$$

where $sort(P_D, N)$ selects the top N workers with the highest corruption probabilities. Based on this framework, the optimal defense strategy is formulated as the minimization problem Q, i.e.,

$$Q: \min_{P_D} \sum_{n=1}^{|O|} sgn(x_{a,n}^{truth} - x_n^{truth}), \tag{5.47a}$$

$$\text{s.t. } U_D = \{u_m | \forall p_m \in sort(P_D, N)\}, \tag{5.47b}$$

$$\widetilde{Z} = \widetilde{X} \backslash \{\widetilde{x}_n^m \mid \widetilde{u}_m \in \widetilde{U} \cap U_D\}, \tag{5.47c}$$

$$\widehat{Z} = \widehat{X} \backslash \{\widehat{y}_n^m \mid \widehat{u}_m \in \widehat{U} \cap U_D\}, \tag{5.47d}$$

where $x_{a,n}^{truth}, \forall o_n \in O$ is the optimal solution of problem P_l using the sensory data $Z = \widetilde{Z} \cup \widehat{Z}$. The constraints in Eqs. (5.47c)–(5.47d) indicate that any data originating from the identified compromised workers is removed from the overall dataset $X = \widetilde{X} \cup \widehat{X}$.

In summary, by casting the defense mechanism as a minimization problem, defenders can effectively single out and eliminate those workers whose contributions most significantly benefit the attack, thereby mitigating the overall impact. This approach is further validated by optimization theory.

5.4.2 *Finding Optimal Countermeasures*

Since the objective in Eq. (5.47a) is not differentiable with respect to P_D, gradient-based methods are unsuitable for solving problem Q. Instead, we utilize the Differential Evolution algorithm [31, 32], a robust swarm-based stochastic search method, to tackle Q. The procedure for determining the optimal defenses is outlined in Algorithm 10, known as the Optimal Defense Algorithm (ODA). Initially, the algorithm performs an initialization step, followed by iterative mutation, crossover, and selection processes, as detailed below.

Algorithm 10 Optimal defense algorithm

Input: Size of population SP; Maximum number of iterations gen_{max}; Amplification factor $F \in [0, 2]$; Crossover factor $CR \in [0, 1]$

Output: Corrupted workers U_D detected by defenders

1: Initialize a population of SP individuals $\mathbf{P}_D^0 = \{P_{D,0}^0, P_{D,1}^0, \cdots, P_{D,SP}^0\}$ with $\forall j \in [0, SP]$, $P_{D,j}^0 = \{p_{j,0}^0, p_{j,1}^0, \cdots, p_{j,|U|}^0\}$ uniformly distribution in range $[0, 1]$;
2: $P_{best}^0 \leftarrow P_{D,0}^0$;
3: **for** $j = 0, 1, 2, \cdots, SP$ **do**
4:     **if** $H(P_{best}^0) > H(P_{D,j}^0)$ **then**
5:         $P_{best}^0 \leftarrow P_{D,j}^0$;
6:     **end if**
7: **end for**
8: **for** $G = 1, 2, 3, \cdots, gen_{max}$ **do**
9:     **for** $j = 0, 1, 2, \cdots, SP$ **do**
10:         $r_1, r_2, r_3 \leftarrow rand([0, SP])$;
11:         $B_j^G \leftarrow P_{D,r_1}^{G-1} + F \cdot (P_{D,r_2}^{G-1} - P_{D,r_3}^{G-1})$;
12:     **end for**
13:     **for** $j = 0, 1, 2, \cdots, SP$ **do**
14:         **for** $i = 0, 1, 2, \cdots, |U|$ **do**
15:             **if** $rand(0, 1) > CR$ **then**
16:                 $c_{j,i}^G \leftarrow p_{j,i}^{G-1}$;
17:             **else**
18:                 $c_{j,i}^G \leftarrow b_{j,i}^G$;
19:             **end if**
20:         **end for**
21:     **end for**
22:     **for** $j = 0, 1, 2, \cdots, SP$ **do**
23:         **if** $H(C_j^G) < H(P_{D,j}^{G-1})$ **then**
24:             $P_{D,j}^G \leftarrow C_j^G$;
25:         **end if**
26:         **if** $H(P_{best}^{G-1}) > H(C_j^G)$ **then**
27:             $P_{best}^G \leftarrow C_j^G$;
28:         **end if**
29:     **end for**
30: **end for**
31: $U_D \leftarrow \{u_m | p_m \in sort(P_{best}^G, N)\}$;

During the initialization phase (lines 1 to 7), the ODA begins by generating an initial population of SP individuals sampled from a uniform distribution (line 1). The most optimal individual from this population is then identified based on the fitness function $H(*)$ (lines 2 to 7). Specifically, the initial population, denoted as

$\mathbf{P}_D^0 = \{P_{D,0}^0, \cdots, P_{D,SP}^0\}$, consists of SP individuals. Each individual $\forall P_{D,j}^0 = [p_{j,0}^0, \cdots, p_{j,|U|}^0]_{|U|}] \in \mathbf{P}_D^0$ represents a potential solution to problem Q, where each probability $p_{j,i}^0 \in P_{D,j}$ is drawn from a uniform distribution $U(0, 1)$. The fitness value of each individual $P_{D,j}^0$ (i.e., the j-th solution in the initial population) is computed using the following procedure:

- Identify the set of corrupted workers U_D corresponding to $P_{D,j}^0$ using Eq. (5.46);
- Remove the sensory data uploaded by U_D from the dataset X, resulting in a refined dataset Z (as described in Eqs. (5.47c)–(5.47d));
- Solve problem P_l using the updated sensory data Z to determine the inferred truths X_a^{truth}, then evaluate the fitness value of $P_{D,j}^0$ based on Eq. (5.47a).

During the mutation step (lines 9–12), a mutant vector $B_j^G = [b_{j,0}^G, \cdots, b_{j,|U|}^G]_{|U|}$ is generated for each individual from the previous generation. This vector is computed as

$$B_j^G = P_{D,r_1}^{G-1} + F \cdot (P_{D,r_2}^{G-1} - P_{D,r_3}^{G-1}), \tag{5.48}$$

where F is amplification factor, P_{D,r_1}^{G-1} serves as the base individual, and the indices r_1, r_2, and r_3 are randomly drawn from the uniform distribution $U(0, SP)$.

To boost population diversity, the crossover stage (lines 13–22) is introduced. During this phase, for each individual $P_{D,j}^{G-1} = [p_{j,0}^{G-1}, \cdots, p_{j,|U|}^{G-1}]_{|U|}$ from generation $G-1$ and its corresponding mutant vector $B_j^G = [b_{j,0}^G, \cdots, b_{j,|U|}^G]_{|U|}$, a trial individual $C_j^G = [c_{j,0}^G, \cdots, c_{j,|U|}^G]_{|U|}$ is created as follows,

$$c_{j,i}^G = \begin{cases} p_{j,i}^{G-1}, & \text{if } rand(0, 1) > CR, \\ b_{j,i}^G, & \text{else}, \end{cases} \tag{5.49}$$

where $CR \in [0, 1]$ is crossover factor.

Next, in the selection phase (lines 23–31), a greedy strategy is used to form the new generation $\mathbf{P}_D^G$. Specifically, if the trial individual C_j^G has a fitness value $H(C_j^G)$ lower than that of its parent $P_{D,j}^{G-1}$ (i.e., $H(C_j^G) < H(P_{D,j}^{G-1})$), it replaces the parent; otherwise, the parent is retained. By iteratively applying mutation, crossover, and selection, the algorithm evolves the population to the final generation $\mathbf{P}_D^{gen_{max}}$. The individual with the lowest fitness value in this final population is then chosen as the optimal solution to problem Q, thereby minimizing Eq. (5.47a) and reducing the attack gain.

5.5 Computational Complexity and Limitations

5.5.1 Computational Complexity of Attacks and Defenses

In what follows, we analyze the computational complexity of both attack and defense mechanisms using big-O (asymptotic) notation. To determine the optimal attack strategy, attackers execute Algorithm 9. Initially, the costs of executing lines 4, 5, and 6 are $O(|O| \cdot |U|)$, $O(|O|)$, and $O(|O|)$, respectively. As a result, the combined complexity from lines 3 through 8 is $O(c \cdot |O| \cdot (|U| + 2))$, where c denotes the number of iterations over lines 4–6. Next, the steps corresponding to lines 9 to 23 have a complexity of $O(3 \cdot |O|)$. Overall, the effort to compute the optimal data poisoning attack is $O(k_{max} \cdot c \cdot (|O| \cdot (|U| + 5)))$, which can be approximated as $\approx O(k_{max} \cdot c \cdot |O| \cdot |U|)$, with k_{max} representing the maximum number of iterations over lines 3–23. For defense, Algorithm 10 provides the optimal countermeasures. The ODA approach starts with an initialization phase, followed by iterative mutation, crossover, and selection. Their respective complexities are $O(SP \cdot (|U| \cdot \log |U| + 2 \cdot |U| \cdot |O| + |U|))$, $O(SP)$, $O(SP \cdot |U|)$, and $O(SP \cdot (|U| \cdot \log |U| + 2 \cdot |U| \cdot |O|))$, respectively. Therefore, the overall complexity for these defenses is $O(gen_{max} \cdot SP \cdot |U| \cdot (\log |U| + |O|))$.

5.5.2 Limitations of Attacks and Defenses

Our findings indicate that LDP exacerbates the susceptibility of crowdsensing systems to data poisoning attacks. Nonetheless, two challenges remain that warrant further investigation. First, the proposed attacks rely on both truth discovery algorithms and LDP protocols. However, because the truth discovery methods are executed on the crowdsensing server, attackers generally lack insight into them. Second, the effectiveness of the proposed attacks depends on accessing a subset of authentic sensory data, which is obtained by compromising normal workers. To address the first issue, we plan to explore methods that enable attackers to remain covert without needing prior knowledge of the truth discovery mechanisms. Furthermore, we propose optimization-based defense strategies to counteract the vulnerabilities introduced by LDP, and our experiments demonstrate that these defenses can successfully detect compromised workers. One potential drawback of our defensive approach is that it assumes defenders are aware of the attackers' objective to distort true data. To overcome this, future research will focus on techniques for identifying corrupted workers without any such prior background.

5.6 Performance Evaluation

In this section, we validate the effectiveness of our proposed data poisoning attacks and corresponding countermeasures.

5.6.1 Simulation Settings

In this part, we outline the experimental setup, including datasets, comparisons of data poisoning attacks, evaluation metrics, and parameter configurations. All experiments are conducted using Python on a laptop equipped with an Intel i7 CPU and 16GB memory.

Datasets For our experiments, we use two real-world datasets: the Weather Sentiment Dataset [33] and the Duchenne Smile Dataset [34]. Below is a brief description of each.

- *Weather Sentiment Dataset [33].* This dataset involves classifying tweets (objects) based on their sentiment into five categories: Neutral (label: 1), Positive (label: 2), Tweet not Related to Weather (label: 3), Can't Tell (label: 4), and Negative (label: 5). It comprises 6,000 records from 110 workers covering 300 tweets, with each record containing a WorkerID, ObjectID, Response, and Ground-Truth. Table 5.1 presents two example records.
- *Duchenne Smile Dataset [34].* In this dataset, the task is to determine whether a smile in a face image (object) is Duchenne (label: 1) or Non-Duchenne (label: 2). The dataset includes 17,729 records from 64 workers across 2,134 face images, with each record detailing a WorkerID, ObjectID, Response, and Ground-Truth. Two sample records are shown in Table 5.2.

Comparisons of Data Poisoning Attacks We assess our proposed optimal data poisoning attack by benchmarking it against four established attacks: RA, MA, OA, and AD [6]. The following is an overview of each method.

Table 5.1 Sample records in the weather sentiment dataset

WorkerID	ObjectID	Response	Ground-truth
81991168	A2N7I0331X72Z7	5	5
79196168	A2HBI8BAE2C5QH	3	2

Table 5.2 Sample records in the duchenne smile dataset

WorkerID	ObjectID	Response	Ground-truth
AODPVTKUSNBUT	2162	1	1
A31R9OP1C70KJF	2162	1	1

- *Random Attack (RA).* In RA, each corrupted worker randomly uploads sensory data from $[k']$, i.e., $\widetilde{x} \sim U(1, k')$ where symbol $U(\cdot)$ represents a uniform distribution. This method does not involve any coordination among the attackers.
- *Maximum utility damage Attack (MA).* This attack begins with attackers using a truth discovery algorithm (specifically CRH) on the normal sensory data from known workers U^{known} to estimate the true values for all $|O|$ objects. For each object o_n, they compile a candidate set of values that differ from the inferred truth, then have all compromised workers randomly choose a value from this set (i.e., $\forall o_n \in O, \widetilde{x}_n \sim U([k']\backslash\{x_n^{truth}\})$). This strategy aims to maximize disruption of the inferred truths while sacrificing stealth.
- *Optimal stealth Attack (OA).* OA involves corrupted workers submitting data that exactly matches the inferred truth obtained from available real sensory data, meaning $\forall o_n \in O, \widetilde{x}_n = x_n^{truth}$. This attack is designed to remain inconspicuous, foregoing significant alterations to the inferred truth.
- *Attack under Disguise (AD) [6].* AD is a state-of-the-art method targeting discrete crowdsensing systems without privacy protection. Here, each attacker attempts to improve their apparent reliability by agreeing with normal workers on objects where the true value is unlikely to change, while intentionally disagreeing on other objects to maximize the impact on the inferred truths.

Comparison of Defenses Our defense strategy is evaluated against the MWA method [8]. Unlike our approach, which deploys dedicated mechanisms on the crowdsensing server to identify compromised workers, the MWA defense is built around a resilient truth discovery algorithm. In particular, MWA updates inferred truths through three steps. First, for each object $o_n \in O$, the server orders workers' responses in ascending order and partitions them into L groups. Next, it calculates an aggregated truth for each group using Eq. (5.16), yielding L distinct results. Finally, the server determines the final truth for object o_n by taking the median of these aggregated values. This process is mathematically formulated as

$$
\begin{aligned}
x_{a,n}^{truth} =& \mathrm{Median}\Bigg(\frac{-(1+d)k \cdot q + 2\sum_{u \in U_{g,n}^1} w_u \cdot x_n^u}{2(p-q) \cdot \sum_{u \in U_{g,n}^1} w_u}, \\
& \dots, \frac{-(1+d)k \cdot q + 2\sum_{u \in U_{g,n}^L} w_u \cdot x_n^u}{2(p-q) \cdot \sum_{u \in U_{g,n}^L} w_u}\Bigg), \forall o_n \in O
\end{aligned}
\tag{5.50}
$$

where $U_{g,n}^i$ denotes the i-th group for object o_n and the union $U_{g,n}^1 \cup \cdots \cup U_{g,n}^L = U$ covers all workers U. Additionally, MWA uses the same worker-weight updating functions as those in LDP-based privacy-preserving truth discovery (i.e., Eqs. (5.17) and (5.18)) to adjust the reliability of the workers.

Metrics To assess the performance of our data poisoning attacks and countermeasures, we employ the following evaluation metrics.

- *The Effectiveness of Our Proposed Data Poisoning Attacks.* We measure the success of a data poisoning attack by comparing the inferred truth before and after the attack using the Average Utility Damage (AUD). Below, we provide its detailed definition.

$$\text{AUD} := \frac{\sum_{o_n \in O} \mathbb{1}_{x_{a,n}^{truth} \neq x_{g,n}^{truth}}}{|O|}, \tag{5.51}$$

where $x_{g,n}^{truth}$ denotes the ground truth data of object o_n, and the indicator function $\mathbb{1}_{x_{a,n}^{truth} \neq x_{g,n}^{truth}}$ is formulated as

$$\mathbb{1}_{x_{a,n}^{truth} \neq x_{g,n}^{truth}} = \begin{cases} 1 & \text{if } x_{a,n}^{truth} \neq x_{g,n}^{truth} \\ 0 & \text{if } x_{a,n}^{truth} = x_{g,n}^{truth} \end{cases}. \tag{5.52}$$

Thus, a higher AUD signifies a more effective attack.
- *The Effectiveness of Our Proposed Countermeasures.* The defensive performance is assessed using AUD (as defined in Eq. (5.51)), where a lower value signifies better protection. In addition, since our defense mechanism functions as a classifier for identifying compromised workers, its efficacy is also measured using AUC (i.e., the area under the ROC curve), with higher AUC values indicating superior performance.

Parameter Setting We detail the experimental parameters for two aspects: the Optimal Data Poisoning Attacks and the Countermeasures.

- *Optimal Data Poisoning Attacks.* As described in Sect. 5.3, two critical parameters influence our attack's performance: ψ representing the percentage of known workers (detailed in Definition 5.2), and φ representing the percentage of corrupted workers (detailed in Definition 5.1). In our setup, attackers have access to the nor sensory data from $|U^{known}| = \psi \times |\widehat{U}|$ normal workers and control $|\widetilde{U}| = \varphi \times |U|$ corrupted workers. Based on empirical findings (detailed in Sect. 5.6.2), we set $\varepsilon = 1.8$, $\psi = 0.3$, and $\varphi = 0.15$. Moreover, the following parameters in Algorithm 9 are configured empirically: initial Lagrange multipliers $\lambda^0 = \mathbf{0}$, $\mu^0 = \mathbf{0}$, and $\nu^0 = \mathbf{0}$, the penalty factor $\sigma_0 = 2$, the threshold for convergence $\eta = 0.01$, the constraint violation constant $\tau = 0.05$, and the constants $\alpha = 0.1$, $\beta = 0.2$, and $\rho = 3$.
- *Countermeasures.* As outlined in Sect. 5.4, defenders are assumed to know the number of corrupted workers. Thus, only a few empirical parameters in Algorithm 10 require setting: the population size SP, maximum number of iterations gen_{max}, amplification factor F, and crossover factor CR. Following [32], we assign $SP = 20$, $gen_{max} = 1000$, $F = 0.5$, and $CR = 0.1$.

5.6.2 Simulation Results

In this part, we present the detailed simulation results of the proposed attacks and defenses.

Attack Evaluation We demonstrate that data poisoning attacks can effectively compromise crowdsensing systems employing LDP-based privacy preservation. Following this, we assess the attack's performance under various parameter configurations. Moreover, Table 5.3 outlines the computational complexity of these attacks, revealing that our proposed method incurs only a modest increase in cost compared to alternative approaches. In setting with a very large number of workers and objects, where $|U| \gg k_{max}, c$ and $|O| \gg k_{max}, c$, the complexity $O(k_{max} \cdot c \cdot |O| \cdot |U|)$ effectively approximates $O(|O| \cdot |U|)$, indicating that the extra computational burden is minimal.

Figures 5.3 and 5.4 demonstrate the effectiveness of our poisoning attack to LDP-based privacy-preserving crowdsensing systems. In Fig. 5.3, for a fixed privacy budget ε, our attack consistently yields a much higher AUD compared to RA, OA, MA, and AD. This advantage arises because those alternative attacks are not tailored to exploit the nuances of the LDP-based truth discovery process, particularly the increased risk from the calibration step inherent in LDP protocols. Moreover, as the privacy budget ε decreases, the amplification of malicious data becomes

Table 5.3 Computational complexity of data poisoning attacks

Attack strategy	Computational complexity
RA	$O(\|O\| \cdot \|U\|)$
MA	$O(\|O\| \cdot \|U\|)$
OA	$O(\|O\| \cdot \|U\|)$
AD	$O(k_{max} \cdot \|O\| \cdot \|U\|)$
Our	$O(k_{max} \cdot c \cdot \|O\| \cdot \|U\|)$

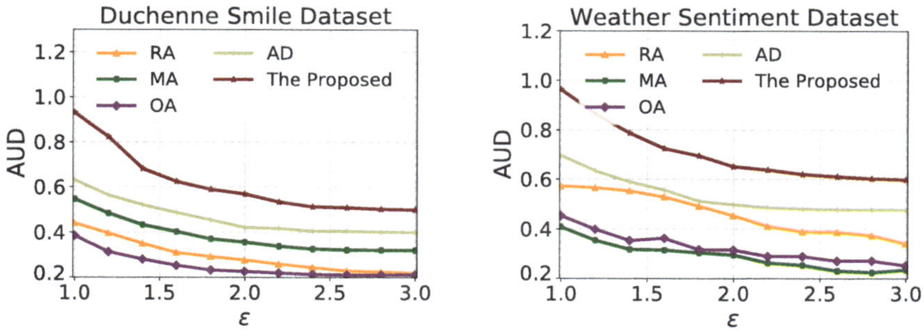

(a) Impact of ε on effectiveness (b) Impact of ε on effectiveness

Fig. 5.3 The effectiveness of data poisoning attacks in both real-world datasets varying with privacy budget ε. (**a**) AUD in the Duchenne smile dataset and (**b**) AUD in the weather sentiment dataset

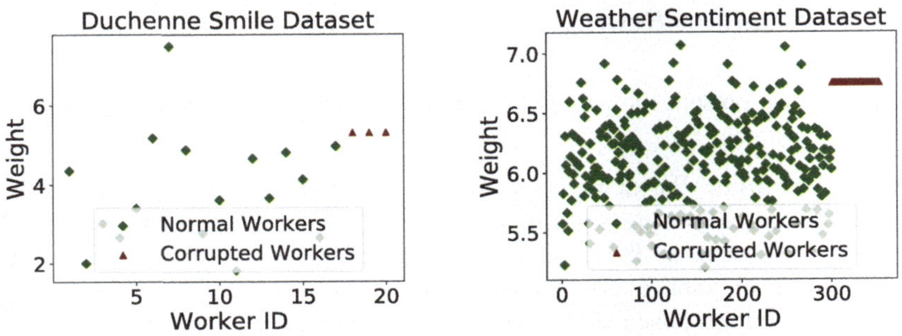

(a) Privacy budget $\varepsilon = 1.8$ (b) Privacy budget $\varepsilon = 1.8$

Fig. 5.4 The weights of workers U under our proposed data poisoning attack against LDP-based privacy-preserving crowdsensing with privacy budget $\varepsilon = 1.8$. (**a**) Worker weights in the Duchenne smile dataset and (**b**) worker weights in the weather sentiment dataset

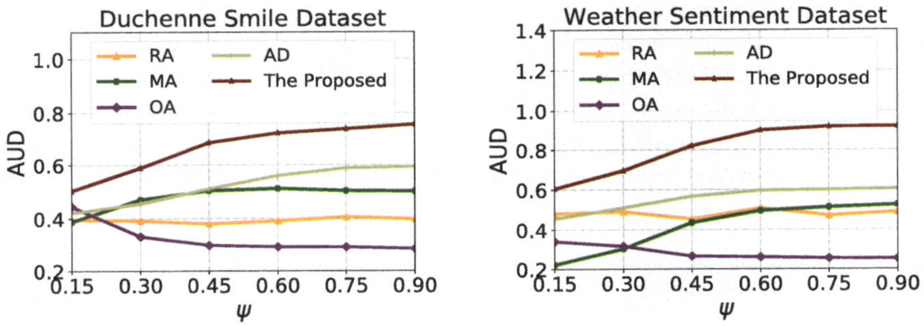

(a) Impact of ψ on effectiveness (b) Impact of ψ on effectiveness

Fig. 5.5 The effectiveness of data poisoning attacks in both real-world datasets varying with parameter ψ (the percentage of known workers). (**a**) AUD in the Duchenne smile dataset and (**b**) AUD in the weather sentiment dataset

more significant, resulting in even higher AUD values, as detailed in Eq. (5.16). Additionally, Fig. 5.4 indicates that our attack circumvents the truth discovery mechanism by leveraging LDP perturbations, which leads to higher weights being assigned to corrupted workers. This uniform weight assignment is a consequence of the consistent manipulation of sensory data (as explained by the transition from optimization problem P_u^1 to P_u^2). In summary, the results in these figures confirm that LDP can inadvertently facilitate the success of data poisoning attacks by enabling them to evade conventional truth discovery methods.

Figure 5.5 illustrates how the AUD varies with the parameter ψ (i.e., the percentage of known workers defined in Definition 5.2) for several data poisoning attacks, namely our proposed method, RA, MA, AD, and OA, across two real-world datasets. Notably, as ψ increases, the AUD values for our method, RA, MA, and AD also rise, whereas OA exhibits a slight decline. This behavior can

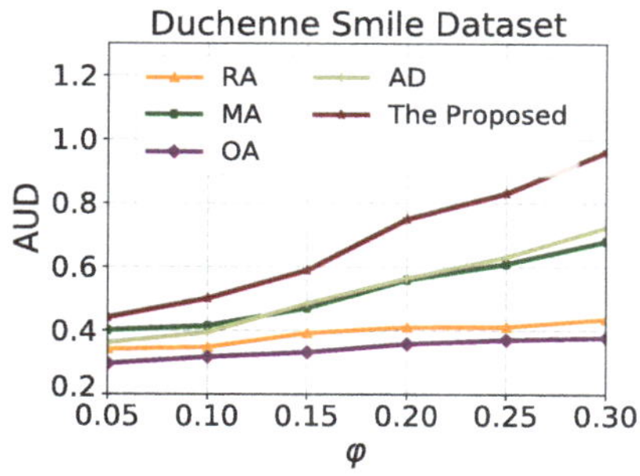

(a) Impact of φ on effectiveness

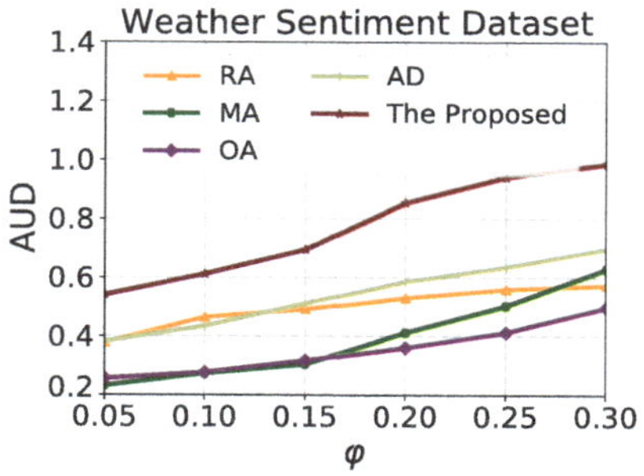

(b) Impact of φ on effectiveness

Fig. 5.6 The effectiveness of data poisoning attacks in both real-world datasets varying with parameter φ (the percentage of corrupted workers). (**a**) AUD in the Duchenne smile dataset and (**b**) AUD in the weather sentiment dataset

be attributed to the fact that a higher ψ allows attackers to access more perturbed sensory data from normal workers, which in turn leads to more precise estimates of the true values (i.e., a smaller difference between the inferred X^{truth} and the actual X_r^{truth}). Although all methods become more stealthy with increased ψ, OA's malicious data remains closer to the true data, hence its decreasing AUD. Moreover, the figure demonstrates that our proposed attack significantly outperforms RA, MA, AD, and OA in terms of AUD, with its advantage growing as ψ increases. This superior performance is due to our attack's optimal stealth strategy for LDP-based privacy-preserving crowdsensing, which exploits the calibration mechanism in LDP protocols to magnify the effect of the malicious data, a factor that the other attacks do not consider.

Figure 5.6 examines how the proportion of corrupted workers (φ, as defined in Definition 5.1) influences attack effectiveness, measured by AUD. The graph indicates that as φ increases, so does the AUD for our proposed data poisoning attack. This trend aligns with our expectations because a higher percentage of compromised workers allows attackers to inject more malicious sensory data, thereby intensifying the attack. Moreover, the figure demonstrates that our method consistently achieves a higher AUD than all the comparison attacks, confirming that by strategically leveraging LDP noise, attackers can effectively undermine truth discovery methods and maximize their impact on the true data.

Defense Evaluation Figure 5.7 compares the performance of our countermeasure against MWA in mitigating the proposed data poisoning attacks. Our approach consistently achieves a lower AUD than MWA under the same privacy budget ε. This enhanced performance is due to its optimization-based mechanism, which effectively detects workers that significantly distort the truth data. Moreover, as the privacy budget ε decreases, the performance gap widens, demonstrating that our countermeasure is particularly adept at reducing the risks posed by LDP.

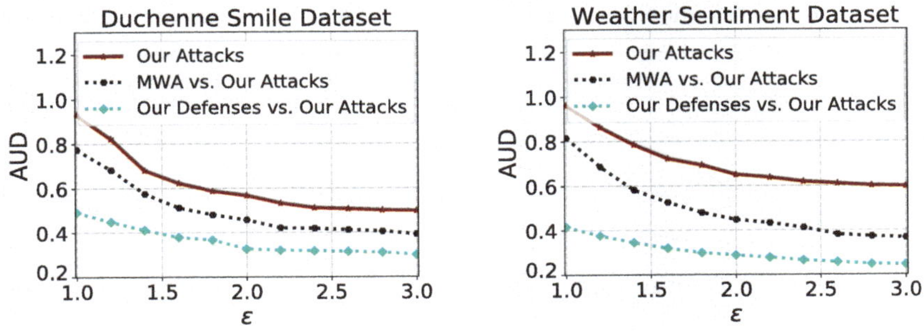

(a) Impact of ε on effectiveness (b) Impact of ε on effectiveness

Fig. 5.7 The effectiveness of defenses (measured by AUD) in both real-world datasets varying with privacy budget ε. (**a**) AUD in the Duchenne smile dataset and (**b**) AUD in the weather sentiment dataset

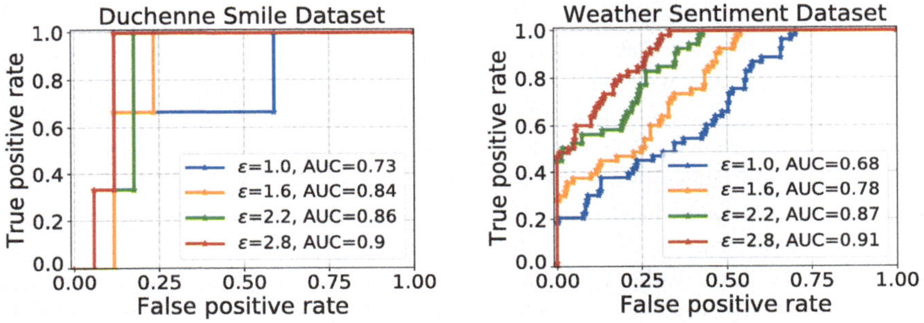

(a) ROC for different ε (b) ROC for different ε

Fig. 5.8 The effectiveness of the proposed countermeasure (measured by AUC) varying with the privacy budget ε. (**a**) ROC in the Duchenne smile dataset and (**b**) ROC in the weather sentiment dataset

Figure 5.8 presents the outcomes of four experimental setups designed to assess how effectively our countermeasure can detect corrupted workers. In these tests, the method is applied under different privacy budgets, $\varepsilon = 1.0, 1.6, 2.2, 2.8$, where the attack patterns are camouflaged within the perturbations. Notably, for the Duchenne smile dataset, an AUC of 0.9 is achieved at $\varepsilon = 2.8$, and even at $\varepsilon = 1.0$, the AUC remains high at 0.73, far exceeding the baseline of 0.5. These results clearly indicate that our countermeasure serves as an effective classifier.

Figure 5.9 illustrates that for the Duchenne smile dataset, the AUC drops markedly when the percentage of known workers (ψ, as defined in Definition 5.2) increases from 0.15 to 0.45, then levels off. In fact, raising φ from 0.15 to 0.90 only leads to about a 0.03% reduction in performance. Similarly, Fig. 5.10 shows a decline in AUC as the percentage of corrupted workers (φ, defined in Definition 5.1) grows. Specifically, in the Duchenne smile dataset, increasing φ from 0.05 to 0.30 results in roughly a 13% decrease in AUC, although the value remains well

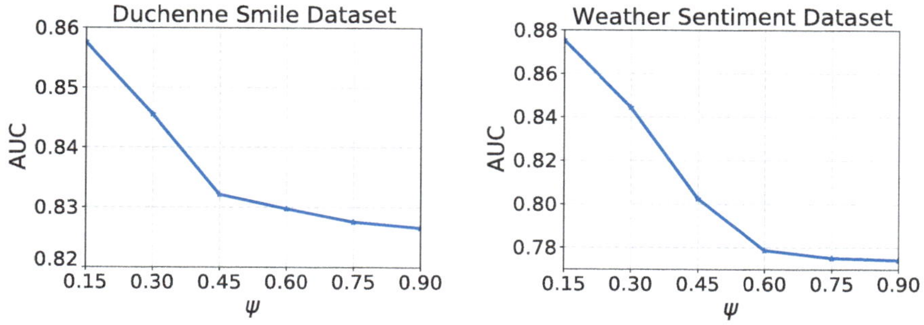

(a) Impact of ψ on AUC (b) Impact of ψ on AUC

Fig. 5.9 The effectiveness of the proposed countermeasure (measured by AUC) varying with parameter ψ (the percentage of known workers). (**a**) AUC in the Duchenne smile dataset and (**b**) AUC in the weather sentiment dataset

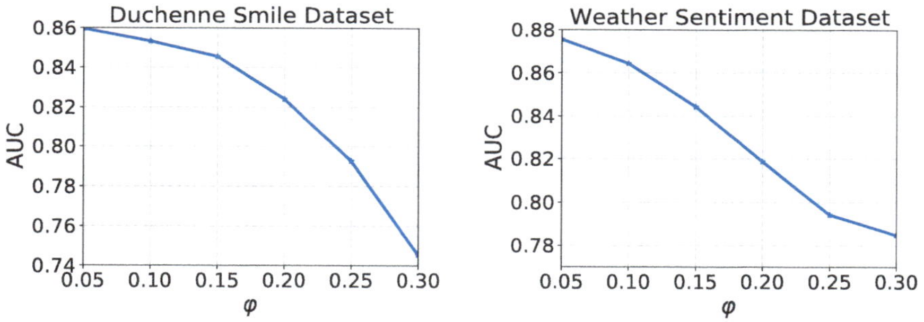

(a) Impact of φ on AUC (b) Impact of φ on AUC

Fig. 5.10 The effectiveness of the proposed countermeasure (measured by AUC) varying with parameter φ (the percentage of corrupted workers). (**a**) AUC in the Duchenne smile dataset and (**b**) AUC in the weather sentiment dataset

Table 5.4 Computational complexity of defenses

Defense strategy	Computational complexity
MWA	$O(\lvert O\rvert \cdot \lvert U\rvert \cdot \log \lvert U\rvert)$
Our	$O(gen_{max} \cdot SP \cdot \lvert U\rvert \cdot (\log \lvert U\rvert + \lvert O\rvert))$

above 0.5 at $\varphi = 0.30$. These modest declines are due to attackers becoming more proficient at concealing their activities (see Figs. 5.5 and 5.6 for details on their disguise capabilities), which makes detection more challenging. Overall, both figures confirm that our countermeasure is robust, as the AUC only experiences slight degradation with increasing ψ and φ.

Table 5.4 compares the computational complexity of our proposed defenses with that of MWA. When there are relatively few participating workers or objects. That is, when $|O| \ll gen_{max} \cdot SP$ or $\log |U| \ll gen_{max} \cdot SP$, our method incurs a slightly higher computational cost than MWA. However, as the number of workers and objects increases, this extra cost becomes negligible. In fact, when $|O| \ll gen_{max} \cdot SP$ or $\log |U| \ll gen_{max} \cdot SP$, the complexities of both approaches converge. Specifically, we have $O(|O| \cdot |U| \cdot \log(|U|)) \approx O(|O|)$ and $O(gen_{max} \cdot SP \cdot |U| \cdot (\log |U| + |O|)) \approx O(|O|)$ when $|O| \gg |U| \gg gen_{max} \cdot SP$, and $O(|O| \cdot |U| \cdot \log(|U|)) \approx O(|U| \cdot \log |U|)$ and $O(gen_{max} \cdot SP \cdot |U| \cdot (\log |U| + |O|)) \approx O(|U| \cdot \log |U|)$ when $\log |U| \gg |O| \gg gen_{max} \cdot SP$.

5.7 Summary and Further Reading

In this chapter, we have proposed the optimal data poisoning attacks and defenses on LDP-based privacy-preserving crowdsensing. The attacks can disrupt the aggregated results of the truth discovery methods by exploiting LDP perturbation, and the defense methods can effectively counter these attacks. Importantly, the proposed attacks expose significant consequences of LDP misuse, which may not only occur in crowdsensing applications but also in other LDP-based applications such as location services. The proposed defenses can be adapted to these applications with minor modifications to defend against similar attacks. For the future work, we will explore stronger data poisoning attacks that exploit personalized noise instead of general noises and design more robust personalized privacy protection solutions.

References

1. Jiang, S., Liu, J., Zhou, Y., Fang, Y.: FVC-dedup: a secure report deduplication scheme in a fog-assisted vehicular crowdsensing system. IEEE Trans. Depend. Secur. Comput. **19**(4), 2727–2740 (2022)
2. Hu, L., Qian, Y., Chen, J., Shi, X., Zhang, J., Mao, S.: Photo crowdsourcing based privacy-protected healthcare. IEEE Trans. Sustain. Comput. **4**(2), 168–177 (2019)
3. Wang, Z., Li, J., Hu, J., Ren, J., Wang, Q., Li, Z., Li, Y.: Towards privacy-driven truthful incentives for mobile crowdsensing under untrusted platform. IEEE Trans. Mobile Comput. **22**(2), 1198–1212 (2023)
4. Cai, C., Zheng, Y., Zhou, A., Wang, C.: Building a secure knowledge marketplace over crowdsensed data streams. IEEE Trans. Depend. Secure Comput. **18**(6), 2601–2616 (2021)
5. Li, Q., Li, Y., Gao, J., Zhao, B., Fan, W., Han, J.: Resolving conflicts in heterogeneous data by truth discovery and source reliability estimation. In: Proceedings of ACM SIGMOD International Conference on Management of Data, pp. 1187–1198 (2014)
6. Miao, C., Li, Q., Su, L., Huai, M., Jiang, W., Gao, J.: Attack under disguise: an intelligent data poisoning attack mechanism in crowdsourcing. In: Proceedings of World Wide Web Conference, pp. 13–22 (2018)
7. Miao, C., Li, Q., Xiao, H., Jiang, W., Huai, M., Su, L.: Towards data poisoning attacks in crowd sensing systems. In: Proceedings of ACM International Symposium on Mobile Ad Hoc Networking and Computing, pp. 111–120 (2018)
8. Fang, M., Sun, M., Li, Q., Gong, N.Z., Tian, J., Liu, J.: Data poisoning attacks and defenses to crowdsourcing systems. In: Proceedings of World Wide Web Conference, pp. 969–980 (2021)

9. Li, Z., Zheng, Z., Guo, S., Guo, B., Xiao, F., Ren, K.: Disguised as privacy: data poisoning attacks against differentially private crowdsensing systems. IEEE Trans. Mobile Comput. **22**(9), 5155–5169 (2023)
10. Li, Y., Xiao, H., Qin, Z., Miao, C., Su, L., Gao, J., Ren, K., Ding, B.: Towards differentially private truth discovery for crowd sensing systems. In: Proceedings of IEEE International Conference on Distributed Computing Systems, pp. 1156–1166 (2020)
11. Huang, C., Liu, D., Yang, A., Lu, R., Shen, X.: Multi-client secure and efficient DPF-based keyword search for cloud storage. IEEE Trans. Depend. Secur. Comput. **21**(1), 353–371 (2023)
12. Zheng, Z., Li, Z., Jiang, H., Zhang, L.Y., Tu, D.: Semantic-aware privacy-preserving online location trajectory data sharing. IEEE Trans. Inf. Forensics Secur. **17**, 2256–2271 (2022)
13. Huang, C., Xue, L., Liu, D., Shen, X., Zhuang, W., Sun, R., Ying, B.: Blockchain-assisted transparent cross-domain authorization and authentication for smart city. IEEE Internet Things J. **9**(18), 17194–17209 (2022)
14. Liu, D., Wu, H., Huang, C., Ni, J., Shen, X.: Blockchain-based credential management for anonymous authentication in sagvn. IEEE J. Sel. Areas Commun. **40**(10), 3104–3116 (2022)
15. Cai, C., Zheng, Y., Du, Y., Qin, Z., Wang, C.: Towards private, robust, and verifiable crowdsensing systems via public blockchains. IEEE Trans. Depend. Secur. Comput. **18**(4), 1893–1907 (2021)
16. Zheng, Z., Li, Z., Li, J., Jiang, H., Li, T., Guo, B.: Utility-aware and privacy-preserving trajectory synthesis model that resists social relationship privacy attacks. ACM Trans. Intell. Syst. Technol. **13**(3), 1–28 (2022)
17. Zheng, Y., Duan, H., Yuan, X., Wang, C.: Privacy-aware and efficient mobile crowdsensing with truth discovery. IEEE Trans. Depend. Secur. Comput. **17**(1), 121–133 (2020)
18. Li, Y., Sun, H., Wang, W.H.: Towards fair truth discovery from biased crowdsourced answers. In: Proceedings of ACM SIGKDD International Conference on Knowledge Discovery & Data Mining, pp. 599–607 (2020)
19. Li, Y., Li, Q., Gao, J., Su, L., Zhao, B., Fan, W., Han, J.: Conflicts to harmony: a framework for resolving conflicts in heterogeneous data by truth discovery. IEEE Trans. Knowl. Data Eng. **28**(8), 1986–1999 (2016)
20. Erlingsson, U., Pihur, V., Korolova, A.: Rappor: randomized aggregatable privacy-preserving ordinal response. In: Proceedings of ACM SIGSAC Conference on Computer and Communications Security, pp. 1054–1067 (2014)
21. Kasiviswanathan, S.P., Lee, H.K., Nissim, K., Raskhodnikova, S., Smith, A.: What can we learn privately? In: Proceedings of IEEE Annual Symposium on Foundations of Computer Science, pp. 531–540 (2008)
22. Duchi, J.C., Jordan, M.I., Wainwright, M.J.: Local privacy and statistical minimax rates. In: Proceedings of IEEE Annual Symposium on Foundations of Computer Science, pp. 429–438 (2013)
23. Zhang, C., Zhu, L., Xu, C., Liu, X., Sharif, K.: Reliable and privacy-preserving truth discovery for mobile crowdsensing systems. IEEE Trans. Depend. Secur. Comput. **18**(3), 1245–1260 (2021)
24. Cao, X., Jia, J., Gong, N.Z.: Data poisoning attacks to local differential privacy protocols. In: Proceedings of USENIX Security Symposium, pp. 947–964 (2021)
25. Cheu, A., Smith, A., Ullman, J.: Manipulation attacks in local differential privacy. In: Proceedings of IEEE Symposium on Security and Privacy, pp. 883–900 (2021)
26. Wu, Y., Cao, X., Jia, J., Gong, N.Z.: Poisoning attacks to local differential privacy protocols for Key-Value data. In: Proceedings of USENIX Security Symposium, pp. 519–536 (2022)
27. Thomas, K., McCoy, D., Grier, C., Kolcz, A., Paxson, V.: Trafficking fraudulent accounts: The role of the underground market in twitter spam and abuse. In: Proceedings of USENIX Security Symposium, pp. 195–210 (2013)
28. Giraldo, J., Cárdenas, A.A., Kantarcioglu, M., Katz, J.: Adversarial classification under differential privacy. In: Proceedings of Annual Network and Distributed System Security Symposium, pp. 1–18 (2020)

29. Warner, S.L.: Randomized response: a survey technique for eliminating evasive answer bias. J. Am. Stat. Assoc. **60**(309), 63–69 (1965)
30. Nie, J., Wang, L., Ye, J.J.: Bilevel polynomial programs and semidefinite relaxation methods. SIAM J. Optim. **27**(3), 1728–1757 (2017)
31. Qin, A.K., Huang, V.L., Suganthan, P.N.: Differential evolution algorithm with strategy adaptation for global numerical optimization. IEEE Trans. Evol. Comput. **13**(2), 398–417 (2009)
32. Storn, R., Price, K.: Differential evolution-a simple and efficient heuristic for global optimization over continuous spaces. J. Global Optim. **11**(4), 341–359 (1997)
33. Venanzi, M., Teacy, W., Rogers, A., Jennings, N.: Weather sentiment - amazon mechanical turk dataset (2015). https://doi.org/10.5258/SOTON/376543
34. Whitehill, J., Wu, T.f., Bergsma, J., Movellan, J., Ruvolo, P.: Whose vote should count more: optimal integration of labels from labelers of unknown expertise. In: Proceedings of the Advances in Neural Information Processing Systems, vol. 22, pp. 1–9 (2009)

Chapter 6
Data Poisoning Attacks and Defenses to CDP-Based Crowdsensing

6.1 Introduction

Crowdsensing is increasingly recognized as an effective means for data collection. Mobile devices equipped with sensors serve as data collectors, capturing a variety of information ranging from environmental conditions to smart transportation insights [1, 2]. In practice, a central server receives tasks from requesters, each outlining specific objectives, and distributes them to selected devices. To minimize individual errors, multiple devices are often assigned to the same task. The server then consolidates the data gathered and returns the aggregated results to the requesters. For instance, a task might involve monitoring specific locations with the goal of assessing noise levels at each site.

However, collected sensor information may contain private details like location and health data [3–9], which raises significant privacy concerns and even deters workers' participation. To address this issue, central differential privacy (CDP) [10] has been broadly adopted in crowdsensing to prevent individual data leakages [4, 11–14]. In this framework, noise is injected into the data prior to sharing, effectively balancing privacy protection with minimal impact on data utility. As a result, the server only processes the altered (noisy) data to compute the aggregated outcomes.

Although privacy is a major issue in crowdsensing, it is not the only threat. Attackers may coerce malicious participants into submitting falsified sensory data for their own gain, a tactic known as data poisoning attacks [15–20]. For example, by disseminating false reports of extreme weather such as a severe storm or heavy rainfall, attackers can distort weather forecasts, leading to poor user experiences and undermining the credibility of weather monitoring crowdsensing systems. This form of attack becomes particularly challenging when truth discovery methods, which aggregate data based on worker reliability, are in place. These methods assess the quality of each worker's contributions, and consistently divergent data results in lower reliability scores, thereby diminishing the influence of those malicious participants in the aggregated outcomes.

Z. Zheng et al., *Perturbation Based Privacy in Crowdsensing*, Wireless Networks,
https://doi.org/10.1007/978-3-031-95052-0_6

Numerous recent studies [15–21] have revealed that truth discovery methods can be significantly undermined by sophisticated data poisoning attacks. In these scenarios, adversarial workers may strategically align with the majority on items that are less likely to change, thereby boosting their reliability scores and exerting greater influence on other tasks. Earlier research targeting malicious behavior has primarily focused on crowdsensing systems where workers directly submit data without the use of CDP, leading to the development of various robust truth discovery techniques [22, 23]. Furthermore, several studies [24–30] have demonstrated that although CDP offers provable privacy guarantees, it can unintentionally facilitate successful data poisoning attacks. These works examine a range of data types, including tabular, key-value, and graph data, within aggregation frameworks but often overlook the challenges that detection mechanisms like truth discovery face against such attacks. In a related study [5], it was shown that malicious workers could evade detection by concealing their actions within the CDP noise. Nonetheless, there is still a limited amount of research on strategies to mitigate the threat posed by attackers hidden in CDP noise.

Unlike previous studies, our work centers on crowdsensing systems that employ central differential privacy, and we aim to develop a robust defense against data poisoning attacks launched by sophisticated attackers. We assume that attackers target the aggregated results produced by truth discovery methods, armed with full knowledge of the defense strategies as well as the deployed CDP protocol and truth discovery technique. Since CDP noise is sampled from the same distribution as genuine sensory data and does not alter the relative differences among data distributions, it is not flagged as malicious by truth discovery approaches [5, 24]. Consequently, attackers can exploit this noise to bypass detection. This scenario presents two major challenges. The first is to *devise an effective defense strategy without knowing the specifics of the poisoning tactic*. The second is to *identify a potent data poisoning strategy that leverages CDP noise to remain undetected*. Additionally, we assume that attackers have access to the raw data collected by some workers.

To overcome these challenges, we introduce a game-theoretic defense framework that enables defenders to develop an optimal countermeasure against potent data poisoning attacks. In this framework, defenders (a.k.a. leaders) choose their defense approach while anticipating that attackers will observe and adapt to it. Specifically, defenders, acting as the game leaders, move first with the goal of reducing harm from adversarial workers. They model their task as a functional optimization problem, seeking to minimize damage by filtering out sensory data from workers identified as malicious through a log-likelihood ratio test. Solving this problem yields the current defense strategy. Subsequently, attackers (a.k.a. followers), taking on the role of followers, launch their offensive with the objective of maximizing disruption to the aggregated outcomes. To remain undetected, they employ a two-pronged stealth tactic. First, they exploit the CDP noise by aligning their attack distribution with that of the perturbed sensory data, effectively bypassing truth discovery methods. Next, using their knowledge of the existing defense measures, they adjust their strategy to reduce the likelihood of triggering the log-likelihood ratio test. As a result, the attackers' maneuvers are formulated as a bi-level optimization problem, where the goal is to maximize the impact on aggregated results under stealth constraints. Solving this problem allows attackers to determine their optimal current strategy. Ultimately, this game-theoretic model captures the dynamic scenario where attackers, fully aware of the defense strategy, continuously refine their tactics based on that knowledge.

Because both the defenders' and attackers' optimization problems are NP-hard, achieving a global minimax solution for the proposed game is extremely challenging. To address this, we introduce the concept of local minimax points along with a key lemma, and we develop an algorithm to compute these points. This algorithm enables us to identify a potent data poisoning attack strategy while simultaneously deriving an effective defense approach. We evaluated our game-theoretic method on two datasets, and the results confirm its effectiveness in countering advanced data poisoning attacks within CDP-based privacy-preserving crowdsensing systems. For example, on the Emotion dataset, the attack gain (as defined in Definition 6.2) decreases by 181.9455 with the Laplace mechanism (or 94.5856 with the Gaussian mechanism) when $\varepsilon = 0.1$. The main contributions of this work are summarized as follows:

- We propose a game-theoretic defense mechanism for CDP-based privacy-preserving crowdsensing that effectively counters sophisticated data poisoning attacks. In our model, the interaction between defenders and attackers is framed as a zero-sum Stackelberg game, where attackers adapt their strategies based on the observed defense. We also introduce an algorithm for computing local minimax points to find an equilibrium in this game.
- We cast the defense strategy as a functional optimization problem aimed at mitigating damage from malicious workers. Since the decision variables are functions and cannot be solved directly, we relax the problem by optimizing the coefficients in basis-function linear combinations via a variable-basis approximation. We then apply a simulated annealing genetic algorithm to solve the relaxed problem and derive the current defense strategy.
- We formulate the attack strategy as a bi-level optimization problem designed to maximize the disruption of truth discovery outputs. This problem is split into an upper-level and a lower-level sub-problem, with the upper level solved analytically using variational methods, enabling attackers to iteratively refine and determine their optimal strategy.

This chapter is structured as follows. Section 6.2 introduces the problem and explains two key concepts: differential privacy and truth discovery methods. Next, Sect. 6.3 models the task of devising an effective defense against potent data poisoning attacks as a zero-sum Stackelberg game. In Sect. 6.4, we detail the development of attack and defense strategies based on this game framework. Section 6.5 then examines the computational complexity of the proposed approaches, with performance evaluations presented in Sect. 6.6. The chapter concludes with a summary in Sect. 6.7.

6.2 Problem Statement and Basic Concepts

6.2.1 Problem Statement

We state the problem addressed in this chapter from the following three aspects: the system model, the defenders and attackers, the attack gain metric, and the problem definition.

System Model A typical crowdsensing system that employs CDP for privacy preservation includes a semi-honest server, several requesters, and a mix of normal and malicious workers. As illustrated in Fig. 6.1, the server, while delivering quality services, may still exhibit curiosity about the private information of the workers. Requesters submit various tasks such as monitoring air quality to the server, which then distributes these tasks among participating workers. The workers use sensors on their mobile devices to complete the assigned tasks and upload the resulting data. To protect their privacy, normal workers perturb their real sensory data using a CDP protocol prior to sharing. Conversely, malicious workers may attempt to disrupt the crowdsensing process by intentionally uploading forged sensory data. After aggregating all the submissions, the server applies a truth discovery method to extract accurate information (a.k.a. the truths), which is then provided back to the requesters as the final aggregated results.

Consider a task that involves a set of objects $O = \{o_1, \cdots, o_{|O|}\}$, where each o_n represents the n-th object. The task is executed by a group of workers $U = \{u_1, \cdots, u_{|U|}\}$, where each u_m denoting the m-th worker. This worker pool

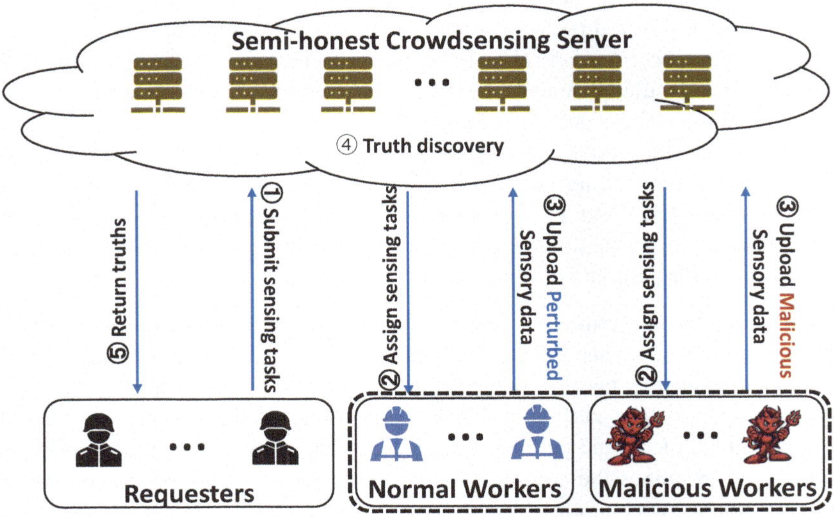

Fig. 6.1 System model

is divided into normal workers $\widehat{U} = \{\widehat{u}_1, \cdots, \widehat{u}_{|\widehat{U}|}\}$ ($\widehat{u}_m$ is m-th normal worker) and malicious workers $\widetilde{U} = \{\widetilde{u}_1, \cdots, \widetilde{u}_{|\widetilde{U}|}\}$ ($\widetilde{u}_m$ denotes m-th malicious worker). The sensory data provided by all workers is represented by $Y = \{y_n^m \mid \forall o_n \in O, \forall u_m \in U\}$, where y_n^m is the data uploaded by the m-th worker for n-th object. This dataset is composed of two parts: the perturbed data $\widehat{Y} = \{\widehat{y}_n^m \mid \forall o_n \in O, \forall \widehat{u}_m \in \widehat{U}\}$ with each $\widehat{y}_n^m$ (and its true counterpart $\widehat{x}_n^m$) coming from a normal worker, and the malicious data $\widetilde{X} = \{\widetilde{x}_n^m \mid \forall o_n \in O, \forall \widetilde{u}_m \in \widetilde{U}\}$ where $\widetilde{x}_n^m$ represents the data from a malicious worker. Subsequently, the server aggregates the combined data $Y = \widehat{Y} \cup \widetilde{X}$ using weight-based truth discovery techniques to estimate the ground truth for each object: $X^{truth} = \{x_n^{truth} \mid \forall o_n \in O\}$. Additionally, the quality or credibility of each worker's contribution is encapsulated in the worker weights W, whose detailed definition is provided below.

Definition 6.1 (Worker Weights) Let $U = \{u_1, \cdots, u_{|U|}\}$ represent all workers, and let their corresponding weights be defined as

$$W = \{w_m \mid \forall u_m \in U\}, \tag{6.1}$$

where each w_m is the weight of the worker u_m. A smaller w_m indicates that the sensory data provided by u_m is of lower quality.

Furthermore, the overall weights W are divided into two groups: one for normal workers and one for malicious workers. Specifically, the normal worker weights are given by $\widehat{W} = \{\widehat{w}_m \mid \forall \widehat{u}_m \in \widehat{U}\}$, where each $\widehat{w}_m$ corresponds to the m-th normal worker. Similarly, the malicious worker weights are defined as $\widetilde{W} = \{\widetilde{w}_m \mid \forall \widetilde{u}_m \in \widetilde{U}\}$ with $\widetilde{w}_m$ denoting the weight for the m-th malicious worker.

Defenders and Attackers In our framework, defenders are deployed on the crowdsensing server with the dual purpose of reducing the adverse effects of malicious workers on truth discovery outcomes while ensuring adherence to the CDP protocol. To achieve this, the server incorporates two detection mechanisms: the conventional truth discovery methods and an auxiliary defense strategy. Although truth discovery techniques are effective at evaluating worker reliability, they are prone to deception by clever attackers who present themselves as trustworthy participants [5, 15–20]. Therefore, it is crucial to develop an additional defensive layer that aids the truth discovery process in uncovering malicious workers concealed within CDP noise.

Furthermore, we analyze a worst-case scenario in which attackers, fully aware of the deployed defense mechanisms and truth discovery algorithms, are capable of coercing certain workers to submit falsified sensory data without altering the behavior of legitimate participants. This insider knowledge enables attackers to execute data poisoning attacks on CDP-based privacy-preserving crowdsensing systems, to maximize discrepancies in the truth discovery outcomes. Such attacks are feasible in real-world applications; for example, attackers might target a real-time navigation system to force it to deliver misleading or dangerous routing information.

Attack Gain Metric We quantify the impact of an attack by measuring the deviation it causes in the ground truths using the Euclidean distance. Let $X_b^{truth} = \{x_{b,n}^{truth} \mid \forall o_n \in O\}$ denote the ground truths before any attack, where each $x_{b,n}^{truth}$ is derived from the perturbed sensory data $\widehat{Y}$. After an attack, the updated ground truths, computed from the complete dataset $Y = \widehat{Y} \cup \widetilde{X}$, are represented by $X_a^{truth} = \{x_{a,n}^{truth} \mid \forall o_n \in O\}$. The deviation for an object o_n is then measured as $(x_{b,n}^{truth} - x_{a,n}^{truth})^2$. However, since the data from normal workers include CDP noise, both X_a^{truth} and X_b^{truth} are random, making this direct measurement imprecise. To mitigate this issue, we consider the expected values $(E[X_b^{truth}] - E[X_a^{truth}])^2$ and define the attack gain metric as follows.

Definition 6.2 (Attack Gain Metric) Let $X_b^{truth} = \{x_{b,n}^{truth} \mid \forall o_n \in O\}$ and $X_a^{truth} = \{x_{a,n}^{truth} \mid \forall o_n \in O\}$ represent the inferred ground truths before and after an attack, respectively. We then define the profit of a data poisoning attack as

$$U(X_b^{truth}, X_a^{truth}) = \sum_{o_n \in O} \left(E[x_{b,n}^{truth}] - E[x_{a,n}^{truth}] \right)^2, \tag{6.2}$$

where $E[x_{b,n}^{truth}]$ and $E[x_{a,n}^{truth}]$ are the expected values of the truths for object o_n before and after attacks, respectively.

Problem Definition Given a scenario where

- CDP-based privacy-preserving crowdsensing systems are under data poisoning attacks, and
- attackers striving to maximize the attack gain $U(X_b^{truth}, X_a^{truth})$.

The goal is to develop a robust defense strategy, denoted by $\mathcal{D}_f$, that minimizes this attack gain by accurately identifying and isolating malicious workers. In designing $\mathcal{D}_f$, it is essential to account for attackers who

- possess access to the genuine sensory data from a subset of normal workers, i.e., $\{x_n^m \mid \forall o_n \in O, \forall u_m \in U^{known}\}$,
- are fully informed about the deployed CDP protocol $\mathcal{A}$ and the truth discovery method $\mathcal{T}_d$, and
- know the details of the defense strategy $\mathcal{D}_f$ employed by the system.

Consequently, advanced attackers execute data poisoning attacks to distort the results of truth discovery methods while evading both the truth discovery process $\mathcal{T}_d$ and the defense mechanism $\mathcal{D}_f$. We denote the attack strategy by $\widetilde{X} \sim \mathcal{F}_a$, where $\mathcal{F}_a\{F_{a,n} \mid \forall o_n \in O\}$ and each $F_{a,n}$ specifies how the malicious sensory data $\widetilde{x}_n^m$ for object o_n is selected. This attack scenario can be formulated as

$$\max_{\widetilde{X} \sim \mathcal{F}_a} U(E[\mathcal{T}_d(\mathcal{A}(\widehat{X}))], E[\mathcal{T}_d(\mathcal{D}_f(\mathcal{A}(\widehat{X}) \cup \widetilde{X}))]), \tag{6.3}$$

where $X_b^{truth} := \mathcal{T}_d(\mathcal{A}(\widehat{X}))$ represents the truths inferred from the perturbed sensory data $\mathcal{A}(\widehat{X})$ before the attack, and $X_a^{truth} := \mathcal{T}_d(\mathcal{D}_f(\mathcal{A}(\widehat{X}) \cup \widetilde{X}))$ denotes the truths after the defense strategy $\mathcal{D}_f$ is applied to the combined data $\mathcal{A}(\widehat{X}) \cup \widetilde{X}$. Here, $\mathcal{T}_d(\mathcal{A}(\widehat{X}))$ implies that the CDP noise, introduced via the perturbation mechanism $\mathcal{A}$, is processed by $\mathcal{T}_d$ to obtain the pre-attack truths. Similarly, $\mathcal{T}_d(\mathcal{D}_f(\mathcal{A}(\widehat{X}) \cup \widetilde{X}))$ signifies that after $\mathcal{D}_f$ filters out malicious contributions from the combined dataset, $\mathcal{T}_d$ infers the post-attack truths. Furthermore, we introduce two essential definitions to quantify the attackers' capabilities and the extent of their background knowledge.

Definition 6.3 (Percentage of Known Workers) Let attackers can access the real sensory data from a subset of normal workers, denoted by U^{known} where $U^{known} \subset \widehat{U}$ (with $\widehat{U}$ representing all normal workers). In other words, attackers are able to illicitly obtain or monitor the real sensory data from these workers. The percentage of such workers is given by

$$pk = |U^{known}|/|\widehat{U}|, \tag{6.4}$$

where $|U^{known}|$ and $|\widehat{U}|$ denote the number of known workers and normal workers, respectively.

Definition 6.4 (Percentage of Malicious Workers) Let attackers control a subset of workers, denoted as $\widetilde{U}$, which is part of the overall participant set U ($\widetilde{U} \subset U$). In other words, attackers have the ability to inject compromised workers into the privacy-preserving crowdsensing system. The percentage of malicious workers is formulated as

$$pm = |\widetilde{U}|/|U|, \tag{6.5}$$

where $|\widetilde{U}|$ represents the number of workers under attacker control, and $|U|$ is the total number of participants.

6.2.2 *Basic Concepts*

In this part, we introduce two fundamental notions: central differential privacy and truth discovery methods.

Central Differential Privacy To prevent workers' private information from leaking, crowdsensing systems frequently utilize perturbation-based privacy protection mechanism, such as central differential privacy (CDP) [10]. Specifically, CDP not only offers provable privacy preservation but also maintains data utility, as outlined below.

Definition 6.5 (Central Differential Privacy) Let $\mathcal{R}$ denote the full range of outputs that the perturbation mechanism $\mathcal{A}$ can produce. For any subset $\mathcal{S} \subset \mathcal{R}$

and any two distinct sensory data points $\widehat{x}^{(1)}$ and $\widehat{x}^{(2)}$, the mechanism $\mathcal{A}$ is said to satisfy (ε, δ)-central differential privacy ((ε, δ)-CDP) if and only if

$$Pr(\mathcal{A}(\widehat{x}^{(1)}) \in \mathcal{S}) \leq e^{\varepsilon} \times Pr(\mathcal{A}(\widehat{x}^{(2)}) \in \mathcal{S}) + \delta. \tag{6.6}$$

When $\delta = 0$, $\mathcal{A}$ is said to achieve ε-central differential privacy (ε-CDP).

Based on Definition 6.5, a smaller (ε, δ) pair implies that outputs from different inputs are more indistinguishable, thereby offering enhanced privacy protection for sensory data. Next, we introduce the notion of sensitive information for each normal worker.

Definition 6.6 (Sensitive Information) Let $\widehat{x}^{m,(1)}$ and $\widehat{x}^{m,(2)}$ are two sensory data uploaded by normal worker $\widehat{u}_m$. The sensitive information of any normal worker $\forall \widehat{u}_m \in \widehat{U}$ is denoted as

$$\Delta_{\widehat{u}_m, p} = \max_{\widehat{x}^{m,(1)}, \widehat{x}^{m,(2)} \in \{\widehat{x}_n^m | \forall o_n \in O\}} \|\widehat{x}^{m,(1)} - \widehat{x}^{m,(2)}\|_p, \tag{6.7}$$

where $\|\widehat{x}^{m,(1)} - \widehat{x}^{m,(2)}\|_p$ is L_p-normal. Because $\|\widehat{x}^{m,(1)} - \widehat{x}^{m,(2)}\|_p = (|\widehat{x}^{m,(1)} - \widehat{x}^{m,(2)}|^p)^{1/p}$, the Eq. (6.7) can be reformulated as

$$\Delta_{\widehat{u}_m} = \max_{\widehat{x}^{m,(1)}, \widehat{x}^{m,(2)} \in \{\widehat{x}_n^m | \forall o_n \in O\}} |\widehat{x}^{m,(1)} - \widehat{x}^{m,(2)}|. \tag{6.8}$$

Laplace and Gaussian mechanisms [31] are widely used to achieve (ε, δ)-CDP. Here, we detail these two mechanisms as follows.

Definition 6.7 (Laplace Mechanism) To achieve ε-CDP, Laplace mechanism adds noise $\xi_n^m \in \mathcal{R}$ selected from $Lap(\Delta_{\widehat{u}_m}/\varepsilon)$ to sensory data $\widehat{x}_n^m$ as follows,

$$\widehat{y}_n^m = \widehat{x}_n^m + \xi_n^m. \tag{6.9}$$

Definition 6.8 (Gaussian Mechanism) The Gaussian mechanism achieves (ε, δ)-CDP by adding noise $\xi_n^m \in \mathcal{R}$ to sensory data $\widehat{x}_n^m$ as follows,

$$\widehat{y}_n^m = \widehat{x}_n^m + \xi_n^m. \tag{6.10}$$

The noise ξ_n^m drawn from $N(0, \sigma_m^2)$, where $\sigma_m \geq \sqrt{2\ln(1.25/\delta)} \cdot \Delta_{\widehat{u}_m}/\varepsilon$.

Truth Discovery Methods In crowdsensing, the sensory data gathered is often inconsistent, which has led to the development of various weight-based truth discovery techniques [32–34] to extract reliable information. These methods work by evaluating the reliability (or weight) of each worker based on their submitted data and then aggregating the data accordingly. In this chapter, we adopt a state-of-the-art approach, Conflict Resolution on Heterogeneous (CRH) [32, 34], to demonstrate how to defend against data poisoning attacks in CDP-based privacy-preserving crowdsensing systems. CRH is formulated as the following optimization problem,

$$\min_{X^{truth}, W} \quad f(X^{truth}, W, Y) = \sum_{u_m \in U} w_m \sum_{o_n \in O} (y_n^m - x_n^{truth})^2, \tag{6.11a}$$

$$\text{s.t.} \quad \sum_{u_m \in U} \exp(-w_m) = 1, \; |U| > 1, \tag{6.11b}$$

where $X^{truth} = \{x_n^{truth} \mid o_n \in O\}$ represents the set of inferred truths for all objects, $W = \{w_m \mid u_m \in U\}$ denotes the weights assigned to the workers, and $Y = \{y_n^m \mid o_n \in O, u_m \in U\}$ is the set of sensory data uploaded by all workers. This optimization problem is solved by iteratively alternating between two key steps.

- *Truths Estimate.* In this step, the inferred truths X^{truth} are derived using the current worker weights W. Specifically, for each object, the estimated truth is calculated as

$$x_n^{truth} = \frac{\sum_{u_m \in U} w_m \cdot y_n^m}{\sum_{u_m \in U} w_m}, \; \forall x_n^{truth} \in X^{truth}. \tag{6.12}$$

  This formulation shows that workers with higher weights have a greater influence on the aggregated result. Consequently, if malicious workers receive elevated weights, their false data can more significantly distort the inferred truths X^{truth}.
- *Worker Weights Estimate.* In this step, using the current truths X^{truth}, the weights W are determined by

$$w_m = \log \frac{\sum_{u_m \in U} \sum_{o_n \in O} (y_n^m - x_n^{truth})^2}{\sum_{o_n \in O} (y_n^m - x_n^{truth})^2}, \; \forall w_m \in W. \tag{6.13}$$

  This formulation indicates that workers whose reported data closely aligns with the consensus are assigned higher weights, while those with larger deviations receive lower weights. As a result, adversarial workers can strategically gain higher weights by matching the normal workers on objects where the true values are difficult to distort, thereby increasing their influence on the overall truth estimates.

Most weight-based truth discovery methods update truth estimates and worker weights iteratively. Although they may differ in the specific update functions used (see Eq. (6.12) for truths and Eq. (6.13) for weights), their core iterative structure is similar. This common framework makes it straightforward to apply the proposed game-based defense strategy to alternative truth discovery methods.

6.3 Formulating Game Model Between Attacks and Defenses

6.3.1 Zero-Sum Stackelberg Game

The system distinguishes malicious workers by evaluating whether a random variable y_n^m originates from the normal behavior hypothesis H_b or from the malicious behavior hypothesis H_a^*. A fundamental result in hypothesis testing shows that the optimal receiver operating characteristic (ROC) curve can be obtained via the log-likelihood ratio test, which compares the true distribution $\mathcal{F}_b = \{F_{b,n} \mid \forall o_n \in O\}$ under H_b with the distribution $\mathcal{F}_a^* = \{F_{a,n}^* \mid \forall o_n \in O\}$ under H_a^*. In practice, this approach is implemented by applying the log-likelihood ratio test, formulated as

$$\mathcal{D}_f := p(u_m) = \frac{\wedge(u_m) - \min(V)}{\max(V) - \min(V)}, \ \forall u_m \in U, \tag{6.14}$$

where

$$\wedge(u_m) = \sum_{o_n \in O} \log \frac{F_{a,n}^*(y_n^m)}{F_{b,n}(y_n^m)}, \ \forall u_m \in U, \tag{6.15}$$

and $V = \{\wedge(u_m) \mid \forall u_m \in U\}$. A worker u_m is then deemed malicious if $p(u_m)$ exceeds a predefined empirical threshold τ.

Since defenders must act before knowing the attackers' true strategy, their assumed attack model $\mathcal{F}_a^*$ differs from the actual attack strategy $\mathcal{F}_a$. We model this interaction as a zero-sum Stackelberg game in which defenders are the leaders and attackers the followers. In this framework, defenders first commit to a defense strategy $\mathcal{D}_f$ (which depends on $\mathcal{F}_a^*$ as defined in Eq. (6.14)) to mitigate the damage from malicious workers. Subsequently, attackers choose their optimal strategy $\mathcal{F}_a$ to maximize their gain given the fixed defense $\mathcal{D}_f$. A complete game strategy is represented by the pair $(\mathcal{F}_a^*, \mathcal{F}_a)$, corresponding to the choices made by defenders and attackers, respectively. Both parties are assumed to be rational, always selecting the strategy that maximizes their utility. This setup is zero-sum because any benefit to the attackers is exactly offset by a loss to the defenders, and vice versa. Consequently, the interaction is formulated as a minimax optimization problem, i.e.,

$$\min_{\mathcal{D}_f (\text{or } \mathcal{F}_a^*)} \max_{\widetilde{X} \sim \mathcal{F}_a} U(E[\mathcal{T}_d(\mathcal{A}(\widehat{X}))], E[\mathcal{T}_d(\mathcal{D}_f(\mathcal{A}(\widehat{X}) \cup \widetilde{X}))]). \tag{6.16}$$

Here, defenders act as the minimizing players and attackers as the maximizing players. Next, we provide detailed definitions of the utilities for both attackers and defenders.

Definition 6.9 (Players' Utilities) Given the attack strategy $\mathcal{F}_a$ and the defense strategy $\mathcal{F}_a^*$. Based on the attack gain metric defined in Definition 6.2, the utilities for the attackers and defenders are given by

$$U_{attackers} = U(X_b^{truth}, X_a^{truth}), \tag{6.17}$$

$$U_{defenders} = -U(X_b^{truth}, X_a^{truth}), \tag{6.18}$$

where $U(X_b^{truth}, X_a^{truth})$ is the attack gain. Here, X_b^{truth} refers to the truths before attacks, determined by aggregating the altered sensory data from trusted workers using truth discovery methods. In contrast, X_a^{truth} represents the true values after an attack and is derived through a three-step process: first, the defense strategy $\mathcal{F}_a^*$ is deployed; second, attackers inject malicious data according to $\mathcal{F}_a$; and third, the resulting compromised sensory data is processed via truth discovery techniques to obtain X_a^{truth}.

6.3.2 Unveiling the Normal Behavior of Workers

As outlined in Sect. 6.3.1, it is crucial to accurately determine the distribution $\mathcal{F}_b = \{F_{b,n} \mid \forall o_n \in O\}$ that characterizes the standard behavior of workers. We now introduce two theorems that describe this normal behavior when using the Laplace and Gaussian mechanisms, respectively.

Theorem 6.1 (Normal Behavior Under Laplace Mechanism) *Let the inferred truths before poisoning attacks is $X_b^{truth} = \{x_{b,n}^{truth} \mid \forall o_n \in O\}$. If the sensory data is added CDP noise by the Laplace mechanism, i.e., $Lap(\Delta_{\widehat{u}_m}/\varepsilon)$, $\forall \widehat{u}_m \in \widehat{U}$, the normal behavior $\mathcal{F}_b = \{F_{b,n} \mid \forall o_n \in O\}$ of workers can be formulated as*

$$F_{b,n}(y) = \frac{\varepsilon \cdot |\widehat{U}|}{2\sum_{\widehat{u}_m \in \widehat{U}} \Delta_{\widehat{u}_m}} e^{-\frac{\varepsilon \cdot |y - x_{b,n}^{truth}| \cdot |\widehat{U}|}{\sum_{\widehat{u}_m \in \widehat{U}} \Delta_{\widehat{u}_m}}}, \forall F_{b,n} \in \mathcal{F}_b, \tag{6.19}$$

where $\Delta_{\widehat{u}_m}$ is the sensitive information of user $\widehat{u}_m$ and defined in Definition 6.6.

Proof To protect normal workers' sensory data privacy, under the Laplace mechanism, real data is added with Laplace noise before sharing. That is, the noise $\xi_n^m \sim Lap(0, \Delta_{\widehat{u}_m}/\varepsilon)$, $\forall \widehat{u}_m \in \widehat{U}$. Next, we calculate the distribution that average noise $\xi_n := \frac{\sum_{\widehat{u}_m \in \widehat{U}} \xi_n^m}{|\widehat{U}|}$ (random variable) follows. Since $\forall \xi_n^m$ follow $Lap(0, \frac{\Delta_{\widehat{u}_m}}{\varepsilon})$, we can estimate the scale parameter of distribution that ξ_n obeys using expectation scale $\frac{\sum_{\widehat{u}_m \in \widehat{U}} \Delta_{\widehat{u}_m}}{\varepsilon \cdot |\widehat{U}|}$. Thus, we have $\xi_n \sim Lap(0, \frac{\sum_{\widehat{u}_m \in \widehat{U}} \Delta_{\widehat{u}_m}}{\varepsilon \cdot |\widehat{U}|})$. As a result, the normal behavior $F_{b,n}(y) = Lap(x_{b,n}^{truth}, \frac{\sum_{\widehat{u}_m \in \widehat{U}} \Delta_{\widehat{u}_m}}{\varepsilon \cdot |\widehat{U}|})$, $\forall F_{b,n} \in \mathcal{F}_b$, i.e., Eq. (6.19) is satisfied. □

Theorem 6.2 (Normal Behavior Under Gaussian Mechanism) *Let the inferred truths before poisoning attacks is $X_b^{truth} = \{x_{b,n}^{truth} \mid \forall o_n \in O\}$. If the sensory data* *is added CDP noise by the Gaussian mechanism, i.e.,* $N(0, \sigma_m^2)$, $\forall \widehat{u}_m \in \widehat{U}$, *the normal behavior* $\mathcal{F}_b = \{F_{b,n} \mid \forall o_n \in O\}$ *of workers can be formulated as*

$$F_{b,n}(y) = \frac{1}{\upsilon_n \cdot \sqrt{2\pi}} e^{-\frac{\left(y - x_{b,n}^{truth}\right)^2}{2\upsilon_n}}, \forall F_{b,n} \in \mathcal{F}_b, \tag{6.20}$$

where υ_n^2 *is the variance of distribution* $F_{b,n}$ *and can be calculated by*

$$\upsilon_n^2 = \frac{\sum_{u_m \in \widehat{U}} \sigma_m^2}{|\widehat{U}|}. \tag{6.21}$$

Proof The normal workers' real data is added with Gaussian noise before sharing, i.e., the noise $\xi_n^m \sim N(0, \sigma_m^2)$, $\forall \widehat{u}_m \in \widehat{U}$. Because of the additivity of the Gaussian distribution, i.e., if $\xi_n^1 \sim N(0, \sigma_1^2)$ and $\xi_n^2 \sim N(0, \sigma_2^2)$ then $\xi_n^1 + \xi_n^2 \sim N(0, \sigma_1^2 + \sigma_2^2)$, we have $\frac{\sum_{\widehat{u}_m \in \widehat{U}} \xi_n^m}{|\widehat{U}|} \sim N(0, \frac{\sum_{\widehat{u}_m \in \widehat{U}} \sigma_m^2}{|\widehat{U}|})$. Thus, the normal behavior $F_{b,n}(y) = N(x_{b,n}^{truth}, \frac{\sum_{\widehat{u}_m \in \widehat{U}} \sigma_m^2}{|\widehat{U}|})$, $\forall F_{b,n} \in \mathcal{F}_b$, i.e., Eq. (6.20) is satisfied. □

6.4 Finding Optimal Data Poisoning Attacks and Defenses

According to previous studies [5, 24], CDP protocols are susceptible to data poisoning, which can even enhance the success of such attacks in crowdsensing systems. In essence, attackers can mimic normal workers by aligning their poisoning strategy with the distribution of perturbed sensor data, a technique referred to as "hiding behind the CDP noise." To counter these attacks, we propose a defense mechanism based on a Stackelberg game framework. In this setup, defenders act first by establishing their defense strategy, and attackers follow by launching poisoning attacks. Specifically, the defense employs an "identify-then-drop" approach: malicious workers are first detected using a log-likelihood ratio test, after which their shared data is discarded. This method allows defenders to formulate their strategy as a functional minimization problem (O_D), aimed at reducing the bias introduced by poisoning (see Sect. 6.4.1). On the other side, we assume that attackers, being well-informed about the CDP protocol and the deployed defense, can exploit CDP noise to bypass truth discovery and even restrict the expected log-likelihood ratio test. Consequently, attackers design their poisoning strategy as a bi-level maximization problem (O_A) with the goal of maximizing attack gain. Thus, the interaction between attackers and defenders is modeled as a zero-sum Stackelberg game, through which we determine the optimal defense strategy against these sophisticated data poisoning attacks (detailed in Sect. 6.4.3).

6.4.1 Defense Strategy for Defenders

The defenders, acting as leaders, take the initiative by employing a log-likelihood ratio test to pinpoint malicious workers. This process is modeled as a minimization problem that aims to reduce the adverse effects of malicious data by fine-tuning the defense strategy $\mathcal{F}_a^*$. Specifically, the formulation is given by

$$(O_D) : \min_{\mathcal{F}_a^*} \sum_{o_n \in O} \left(E[x_{b,n}^{truth}] - E[x_{a,n}^{truth}] \right)^2, \tag{6.22a}$$

$$\text{s.t. } U_D = \{u_m \mid p(u_m) > \tau, \forall u_m \in U\}, \tag{6.22b}$$

$$Z = \{z_n^m = y_n^m \mid \forall u_m \in U \backslash U_D, \forall o_n \in O\}, \tag{6.22c}$$

$$E[x_{b,n}^{truth}] = \int_{-\infty}^{+\infty} y \cdot F_{b,n}(y)dy, \tag{6.22d}$$

$$E[x_{a,n}^{truth}] = \int_{-\infty}^{+\infty} y \cdot F_{a,n}^*(y)dy, \tag{6.22e}$$

$$\{X_a^{truth}, W^*\} := \arg\min_{X_a^{truth}, W^*} f(X_a^{truth}, W^*, Z), \tag{6.22f}$$

where the probabilities $p(u_m)$ for each $u_m \in U$ are determined by the defense strategy $\mathcal{F}_a^*$ for each $u_m \in U$ are determined by the defense strategy $\mathcal{F}_a^*$ and are computed according to Eqs. (6.14) and (6.15). The objective in (6.22a) reflects the defenders' goal of maximizing their utility by minimizing the squared difference between the expected truths before and after an attack. Constraints (6.22b) and (6.22c) implement an "identify-then-drop" tactic, where malicious workers are flagged using the log-likelihood ratio test and their data is subsequently removed. Constraints (6.22d) and (6.22e) define the expected values of the true data before and after an attack, respectively. Lastly, constraint (6.22f) encapsulates the truth discovery process (as detailed in Eqs. (6.11a) and (6.11b)), with $W^* = \{w_m \mid u_m \in U \backslash U_D\}$ representing the weights assigned to workers after suspicious data has been excluded.

Because the decision variable $\mathcal{F}_a^*$ consists of a set of functions, the optimization problem (O_D) is a functional one and cannot be solved directly [35]. To overcome this challenge, we utilize variable-basis approximation techniques, which represent the decision variable as a linear combination of predetermined basis functions. This approach transforms the problem from optimizing over functions to optimizing over the coefficients of these combinations [36]. In our method, we choose Hermite polynomials as the basis functions because they form a well-known orthogonal polynomial sequence that is effective for approximating functions over $(-\infty, +\infty))$ [37]. These polynomials satisfy the recurrence relation $H_{e_{r+1}}(x) = x \cdot H_{e_r}(x) - \frac{d}{dx} H_{e_r}(x)$ $(H_{e_r}(x)$, where $H_{e_r}(x)$ denotes the r-th Hermite polynomial. The first four Hermite polynomials are given by $H_{e_0}(x) = 1$, $H_{e_1}(x) = x$,

$H_{e_2}(x) = x^2 - 1$, and $H_{e_3}(x) = x^3 - 3x$. We approximate $\mathcal{F}_a^*$ by expanding it in terms of these Hermite polynomials up to order r, i.e.,

$$\begin{aligned} F_{a,n}^*(x) =& \psi_{n,0}^* + \psi_{n,1}^* \cdot H_{e_1}(x) + \psi_{n,2}^* \cdot H_{e_2}(x) \\ &+ \cdots + \psi_{n,r-1}^* \cdot H_{e_{r-1}}(x),\ \forall o_n \in O, \end{aligned} \tag{6.23}$$

with the coefficient vector $\Psi_n^* = [\psi_{n,0}^*, \psi_{n,1}^*, \cdots, \psi_{n,r-1}^*]_{1\times r}$. Since the basis functions $H_{e_r}(x)$ are predetermined, the original optimization over the function set $\mathcal{F}_a^* = \{F_{a,n}^* \mid \forall o_n \in O\}$ is converted into an optimization over the coefficient matrix $\Phi^* = [\Psi_0^*, \Psi_1^*, \cdots, \Psi_{|O|}^*]_{|O|\times r}^T$. This relaxation simplifies the original problem (O_D) into a new form, denoted by (O_D^1), which can then be effectively solved using numerical optimization algorithms.

Algorithm 11 Defense strategy algorithm

Input: $Y = \{y_n^m \mid \forall o_n \in O, \forall u_m \in U\}$: the collected sensory data set; $\mathcal{F}_b = \{F_{b,n} \mid \forall o_n \in O\}$: the distribution followed by the perturbed sensory data; τ: the empirical parameter utilized in log-likelihood ratio test; SN: the population size; G_{max}: the maximum number of generations; p_c: the crossover probability; p_m: the mutation probability; T_{max}, T_{min}: the initial and final temperatures; β: the cooling factor;

Output: $\Phi^* = [\Psi_0^*, \cdots, \Psi_{|O|}^*]_{|O|\times r}^T$: the coefficient matrix of rth-order Hermite polynomials; U_D: the detected malicious workers set.

$\{\Phi_0^{*,0}, \cdots, \Phi_{SN}^{*,0}\} \leftarrow$ Initialize a population with SN individuals utilizing the initialization operation;

Evaluate the fitness of 0-th population $\{\Phi_0^{*,0}, \cdots, \Phi_{SN}^{*,0}\}$ utilizing the fitness evaluation operation;

$g \leftarrow 0$;

while $g \leq G_{max}$ and $T_{max} > T_{min}$ **do**

  $\{\Theta_0^{*,g}, \cdots, \Theta_{SN}^{*,g}\} \leftarrow$ Generate offspring by selecting parent individuals from g-th population and then performing crossover operation;

  $\{\Xi_0^{*,g}, \cdots, \Xi_{SN}^{*,g}\} \leftarrow$ Generate mutated offspring by performing mutation operation on each $\Theta_k^{*,g}$;

  Evaluate the fitness of each offspring in $\{\Xi_0^{*,g}, \cdots, \Xi_{SN}^{*,g}\}$ utilizing the fitness evaluation operation;

  $\{\Phi_0^{*,g+1}, \cdots, \Phi_{SN}^{*,g+1}\} \leftarrow$ Select SN individuals from $\{\Xi_0^{*,g}, \cdots, \Xi_{SN}^{*,g}\} \cup \{\Phi_0^{*,g}, \cdots, \Phi_{SN}^{*,g}\}$ utilizing the simulated annealing operation;

  $T_{max} \leftarrow \beta \cdot T_{max}$;

  $g \leftarrow g + 1$;

end while

$\Phi^* \leftarrow$ optimal individual in the g-th population;

Since the optimization problem (O_D^1) involves a high-dimensional decision variable in the form of a $|O| \times r$ matrix, we adopt the simulated annealing genetic algorithm (SAGA) [38] to efficiently solve it. The details of this approach are provided in Algorithm 11. SAGA is a hybrid metaheuristic that integrates two powerful optimization techniques: (1) Simulated Annealing (SA) [39], which is a single-solution-based algorithm designed to escape local optima by probabilistically accepting worse solutions; (2) Genetic Algorithm (GA) [40], a population-based approach that enhances search diversity through crossover and mutation operations. By combining these methods, SAGA effectively explores the large search space and improves the likelihood of finding the global optimum. The algorithm follows an iterative process consisting of initialization, crossover, mutation, fitness evaluation, and simulated annealing, each of which is described in detail below.

In the initialization operations, we generate an initial population of SN individuals using a normal distribution. The initial (or 0-th) population is denoted as $\{\Phi_i^{*,0} \mid \forall i \in [0, SN]\}$, where $\Phi_i^{*,0} = [\Psi_{i,0}^{*,0}, \Psi_{i,1}^{*,0}, \cdots, \Psi_{i,|O|}^{*,0}]_{|O|\times r}^T$ is the i-th individual (or solution candidate) and $\Psi_{i,n}^{*,0} = [\psi_{i,n,0}^{*,0}, \psi_{i,n,1}^{*,0}, \cdots, \psi_{i,n,r-1}^{*,0}]_{1\times r}$ is the coefficient vector candidate of the defense strategy $F_{a,n}^*(x)$. Particularly, the coefficient candidate $\forall \psi_{i,n,j}^{*,0} \in \Psi_{i,n}^{*,0} \in \Phi_i^{*,0}$ drawn from $N(\mu, \sigma^2)$, i.e., $\psi_{i,n,j}^{*,0} \sim N(\mu, \sigma^2)$, where μ and σ^2 represent expectation and variance, respectively.

In the fitness evaluation operation, we evaluate the fitness of each individual in g-th population $\{\Phi_0^{*,g}, \cdots, \Phi_{SN}^{*,g}\}$ $(0 \le g \le G_{max})$. The fitness of $\forall \Phi_i^{*,g}$ (i-th individual in g-th population) can be calculated by

$$Fit(\Phi_i^{*,g}) = -\sum_{o_n \in O} (E[x_{b,n}^{truth}] - E[x_{a,n}^{truth}])^2, \forall i \in [0, SN], \tag{6.24}$$

where $E[x_{b,n}^{truth}]$ is a fixed value and $E[x_{a,n}^{truth}]$ is dependent on $\Phi_i^{*,g}$. Specifically, we can obtain the defense strategy $\mathcal{F}_a^*$ by Eq. (6.23) according to the coefficient matrix $\Phi_i^{*,g}$ (i.e., rth-order Hermite polynomials approximation), secondly identify malicious workers through the log-likelihood ratio test (i.e., constraint (6.22b)), thirdly delete the sensory data uploaded by the identified worker U_D from the collected data (i.e., constraint (6.22c)), fourthly infer the truths $x_{a,n}^{truth}$ from the deleted data $Z = \{z_n^m \mid \forall u_m \in U \backslash U_D, \forall o_n \in O\}$ (i.e., constraint (6.22f)), and finally calculate the expectation $E[x_{a,n}^{truth}]$ by constraint (6.22e). As detailed in Sect. 6.2.2, the expectation $x_{a,n}^{truth}$ can be inferred (or the sub-problem (6.22f) can be solved) by iterating the following two steps:

- Fixing the weights $W = \{w_m \mid \forall u_m \in U \backslash U_D\}$, the truths $X_a^{truth} = \{x_{a,n}^{truth} \mid \forall o_n \in O\}$ are calculated by

$$x_{a,n}^{truth} = \frac{\sum_{u_m \in U \backslash U_D} w_m \cdot z_n^m}{\sum_{u_m \in U \backslash U_D} w_m}; \tag{6.25}$$

- After obtaining the truths X_a^{truth}, the weights $W = \{w_m \mid \forall u_m \in U \backslash U_D\}$ are estimated by

$$w_m = \log \frac{\sum_{u_m \in U \backslash U_D} \sum_{o_n \in O} (z_n^m - x_{a,n}^{truth})^2}{\sum_{o_n \in O} (z_n^m - x_{a,n}^{truth})^2}. \tag{6.26}$$

In the crossover operations, we utilize the tournament selection method [41] to select two-parent individuals from the g-th population $\{\Phi_0^{*,g}, \cdots, \Phi_{SN}^{*,g}\}$. Given two-parent individuals $\Phi_i^{*,g} = [\Psi_{i,0}^{*,g}, \cdots, \Psi_{i,|O|}^{*,g}]$ $(i \in [0, SN])$ and $\Phi_j^{*,g} = [\Psi_{j,0}^{*,g}, \cdots, \Psi_{j,|O|}^{*,g}]$ $(j \in [0, SN])$, the crossover operation is performed between them when a randomly generated number ω_c, drawn from a uniform distribution $U(0, 1)$, satisfies $\omega_c < p_c$. Particularly, we randomly select the position vector $\mathcal{I}_c = [I_c^0, I_c^1, \cdots, I_c^{|O|}]$ ($I_c^k \in [0, r-1]$ represents the k-th crossover position between $\Phi_i^{*,g}$ and $\Phi_j^{*,g}$), and then generate offspring individuals by swapping elements (or coefficients) after the designated positions. That is, by swapping the elements after I_c^k $(\forall I_c^k \in \mathcal{I}_c)$ between $\Psi_{i,k}^{*,g} = [\psi_{i,k,0}^{*,g}, \cdots, \psi_{i,k,I_c^k}^{*,g}, \cdots, \psi_{i,k,r-1}^{*,g}]$ $(\forall \Psi_{i,k}^{*,g} \in \Phi_i^{*,g})$ and $\Psi_{j,k}^{*,g} = [\psi_{j,k,0}^{*,g}, \cdots, \psi_{j,k,I_c^k}^{*,g}, \cdots, \psi_{j,k,r-1}^{*,g}]$ $(\forall \Psi_{j,k}^{*,g} \in \Phi_i^{*,g})$, we can obtain $\Gamma_{i,k}^{*,g} = [\psi_{i,k,0}^{*,g}, \cdots, \psi_{i,k,I_c^k-1}^{*,g}, \psi_{j,k,I_c^k}^{*,g}, \cdots, \psi_{j,k,r-1}^{*,g}]$ and $\Gamma_{j,k}^{*,g} = [\psi_{j,k,0}^{*,g}, \cdots, \psi_{j,k,I_c^k-1}^{*,g}, \psi_{i,k,I_c^k}^{*,g}, \cdots, \psi_{i,k,r-1}^{*,g}]$. After the crossover between $\Phi_i^{*,g}$ and $\Phi_j^{*,g}$, we can obtain the offspring individuals $\Theta_i^{*,g} = [\Gamma_{i,0}^{*,g}, \Gamma_{i,1}^{*,g}, \cdots, \Gamma_{i,r-1}^{*,g}]$ and $\Theta_j^{*,g} = [\Gamma_{j,0}^{*,g}, \Gamma_{j,1}^{*,g}, \cdots, \Gamma_{j,r-1}^{*,g}]$. In addition, we denote the g-th population after crossover as $\{\Theta_0^{*,g}, \Theta_1^{*,g}, \cdots, \Theta_{SN}^{*,g}\}$.

In the mutation operation, we perform the mutation operation on the offspring population $\{\Theta_0^{*,g}, \Theta_1^{*,g}, \cdots, \Theta_{SN}^{*,g}\}$. If the randomly selected number ω_m, drawn from a uniform distribution $U(0, 1)$, is less than the mutation probability p_m, the offspring individual $\Theta_i^{*,g} = [\Gamma_{i,0}^{*,g}, \Gamma_{i,1}^{*,g}, \cdots, \Gamma_{i,r-1}^{*,g}]$ $(\forall i \in [0, SN])$ undergoes mutation to produce $\Xi_i^{*,g}$. Specifically, we randomly choose the position vector $\mathcal{I}_m = [I_m^0, I_m^1, \cdots, I_m^{|O|}]$ ($I_m^k \in [0, r-1]$ denotes the k-th mutation position for $\Theta_i^{*,g}$). Then, we choose a random number $\chi_{i,k,I_m^k}^{*,g}$ $(\forall I_m^k \in \mathcal{I}_m)$ from a normal distribution $N(\mu, \sigma^2)$, i.e., $\chi_{i,k,I_m^k}^{*,g} \sim N(\mu, \sigma^2)$, to replace the I_m^k-th element in the $\Gamma_{i,k}^{*,g} = [\psi_{i,k,0}^{*,g}, \cdots, \psi_{i,k,I_m^k}^{*,g}, \cdots, \psi_{i,k,r-1}^{*,g}]$ $(\forall \Gamma_{i,k}^{*,g} \in \Theta_i^{*,g})$, i.e., $\Gamma_{i,k}^{*,g} = [\psi_{i,k,0}^{*,g}, \cdots, \chi_{i,k,I_m^k}^{*,g}, \cdots, \psi_{i,k,r-1}^{*,g}]$. As a result, we can obtain the mutated population $\Xi_i^{*,g} = [\Gamma_{i,0}^{*,g}, \Gamma_{i,1}^{*,g}, \cdots, \Gamma_{i,r-1}^{*,g}]$.

In the simulated annealing operation, we select SN individuals from both the parents population (g-th population $\{\Phi_0^{*,g}, \cdots, \Phi_{SN}^{*,g}\}$) and the offspring population (g-th population $\{\Xi_0^{*,g}, \cdots, \Xi_{SN}^{*,g}\}$) as the next generation population ($g+1$-th population $\{\Phi_0^{*,g+1}, \cdots, \Phi_0^{*,g+1}\}$). Particularly, given parent $\Phi_k^{*,g}$ $(\forall k \in [0, SN])$ and offspring $\Xi_k^{*,g}$ $(\forall k \in [0, SN])$, if offspring $\Xi_k^{*,g}$ is better (i.e., $Fit(\Xi_k^{*,g}) > Fit(\Phi_k^{*,g})$), remain offspring $\Xi_k^{*,g}$ instead of parent $\Phi_k^{*,g}$; else if parent $\Phi_k^{*,g}$ is better (i.e., $Fit(\Phi_k^{*,g}) > Fit(\Xi_k^{*,g})$), still remain offspring $\Xi_k^{*,g}$ with a certain probability. That is, the offspring $\Xi_k^{*,g}$ is remained when $\exp(-(Fit(\Phi_k^{*,g}) - Fit(\Xi_k^{*,g}))/T_{max}) \geq \omega_{sa}$ is holds (ω_{sa} is a random number drawn from a uniform distribution $U(0, 1)$).

Algorithm 12 Attack strategy algorithm

Input: $Y = \{y_n^m \mid \forall o_n \in O, \forall u_m \in U\}$: the collected sensory data set; $\mathcal{F}_b = \{F_{b,n} \mid \forall o_n \in O\}$: the distribution followed by the perturbed sensory data; $\mathcal{F}_a^* = \{F_{a,n}^* \mid \forall o_n \in O\}$: the current defense strategy; κ: the threshold for convergence;

Output: $\mathcal{F}_a = \{F_{a,n} \mid \forall o_n \in O\}$: the current attack strategy.

$\mathcal{F}_a = \{F_{a,n} \mid \forall o_n \in O\} \leftarrow$ The attack strategy is initialized to normal behavior, i.e., $F_{a,n} = F_{b,n}, \forall o_n \in O$;

$i \leftarrow 0$;

$\{X_a^{truth,i}, W\} \leftarrow$ Solve the lower-level problem Eq. (6.29g) using current attack strategy $\mathcal{F}_a$ by iterating Eq. (6.55) and Eqs. (6.56)–(6.57);

while True **do**

  $\mathcal{F}_a = \{F_{a,n} \mid \forall o_n \in O\} \leftarrow$ Solve the upper-level problem (O_A^U) using current worker weights W by Theorem 6.3;

  $\{X_a^{truth,i+1}, W\} \leftarrow$ Solve the lower-level problem Eq. (6.29g) using current attack strategy $\mathcal{F}_a$ by iterating Eq. (6.55) and Eqs.(6.56)-(6.57);

  Calculate the expectations $E[X_a^{truth,i+1}]$ and $E[X_a^{truth,i}]$ by constraints (6.29e) and (6.29f);

  if $|E[X_a^{truth,i+1}] - E[X_a^{truth,i}]| < \kappa$ **then**

    break;

  end if

  $i \leftarrow i + 1$;

end while

6.4.2 *Attack Strategy for Attackers*

Defenders, acting as leaders, start by deploying their defense mechanism $\mathcal{F}_a^*$. In response, attackers, as followers, launch their data poisoning attacks. Because both the defense strategy $\mathcal{F}_a^*$ and the truth discovery method are publicly known, attackers can conceal their actions effectively. Initially, attackers leverage differential privacy (DP) noise to mask their malicious behavior, thereby evading detection by the truth discovery process. Specifically, they estimate the distribution of the perturbed sensory data, where $\widehat{y}_n^m \sim F_{b,n}$ for every $o_n \in O\}$, and then adjust their attack strategy $\mathcal{F}_a = \{F_{a,n} \mid \forall o_n \in O\}$ to closely resemble the estimated distribution $\mathcal{F}_b = \{F_{b,n} \mid \forall o_n \in O\}$. The similarity between the attack and normal distributions is measured using the Kullback-Leibler (KL) divergence [42]:

$$D_{KL}(F_{a,n} \| F_{b,n}) = \int_{-\infty}^{+\infty} F_{a,n}(x) \cdot \log \frac{F_{a,n}(x)}{F_{b,n}(x)} dx. \tag{6.27}$$

Moreover, attackers can manipulate the expected outcome of the log-likelihood ratio test, which used by defenders to detect anomalies, to further avoid being flagged. When defenders implement $\mathcal{F}_a^*$, the expectation is given by

$$E[\wedge] = \sum_{o_n \in O} \int_{-\infty}^{+\infty} F_{a,n}(y) \cdot \log \frac{F_{a,n}^*(y)}{F_{b,n}(y)} dy. \tag{6.28}$$

Based on the discussion above, when the defense strategy $\mathcal{F}_a^*$ is fixed, the attack strategy can be formulated as the following maximization problem,

$$(O_A): \max_{\tilde{X} \sim \mathcal{F}_a} \sum_{o_n \in O} \left(E[x_{b,n}^{truth}] - E[x_{a,n}^{truth}] \right)^2, \tag{6.29a}$$

$$\text{s.t.} \sum_{o_n \in O} D_{KL}(F_{a,n} \| F_{b,n}) \leq \eta, \tag{6.29b}$$

$$E[\wedge] \leq \zeta, \tag{6.29c}$$

$$\int_{-\infty}^{+\infty} F_{a,n}(y) dy = 1, \ \forall o_n \in O, \tag{6.29d}$$

$$E[x_{b,n}^{truth}] = \int_{-\infty}^{+\infty} y \cdot F_{b,n}(y) dy, \tag{6.29e}$$

$$E[x_{a,n}^{truth}] = \int_{-\infty}^{+\infty} y \cdot F_{a,n}(y) dy, \tag{6.29f}$$

$$\{X_a^{truth}, W\} := \arg\min_{X_a^{truth}, W} f(X_a^{truth}, W, Z), \tag{6.29g}$$

where the threshold η is the disguise level hidden behind DP noise, and the threshold ζ dictates the stealth level of evading defenders. A smaller value of η or ζ corresponds to a lower probability of being detected by truth discovery methods or defenders. Specifically, the objective function aims to maximize the attackers' utility $U_{attackers}$ as defined in Definition 6.9). The constraints ensure that the attack remains undetected: the first and second constraints are designed to bypass truth discovery methods and defenders, respectively, while the third to fifth constraints enforce the expectations of the random variables $x_{b,n}^{truth}$ and $x_{a,n}^{truth}$. Lastly, the sixth constraint defines the lower-level optimization problem representing the truth discovery method.

The optimization problem (O_A) is formulated as a bi-level problem [43] and can be divided into two parts: a lower-level sub-problem given by Eq. (6.29g) and an upper-level sub-problem, denoted as (O_A^U). These components are addressed in an alternating manner. In the following, we provide a detailed explanation of how to solve both (O_A^U) and the lower-level sub-problem in Eq. (6.29g).

Upper-Level Sub-problem The upper-level sub-problem (O_A^U) of problem (O_A) is formulated as

$$(O_A^U): \max_{\widetilde{X}\sim\mathcal{F}_a} \sum_{o_n\in O} \left(E[x_{b,n}^{truth}] - E[x_{a,n}^{truth}]\right)^2, \tag{6.30a}$$

$$\text{s.t.} \sum_{o_n\in O} D_{KL}(F_{a,n}\|F_{b,n}) \le \eta, \tag{6.30b}$$

$$E[\wedge] \le \zeta, \tag{6.30c}$$

$$\int_{-\infty}^{+\infty} F_{a,n}(y)dy = 1,\ \forall o_n \in O, \tag{6.30d}$$

$$E[x_{b,n}^{truth}] = \int_{-\infty}^{+\infty} y\cdot F_{b,n}(y)dy, \tag{6.30e}$$

$$E[x_{a,n}^{truth}] = \int_{-\infty}^{+\infty} y\cdot F_{a,n}(y)dy, \tag{6.30f}$$

$$x_{a,n}^{truth} = \frac{\sum_{\widehat{u}_m\in\widehat{U}} \widehat{w}_m\cdot\widehat{x}_n^m + \sum_{\widetilde{u}_m\in\widetilde{U}} \widetilde{w}_m\cdot\widetilde{x}_n^m}{\sum_{\widehat{u}_m\in\widehat{U}} \widehat{w}_m + \sum_{\widetilde{u}_m\in\widetilde{U}} \widetilde{w}_m}, \tag{6.30g}$$

where the truths $x_{a,n}^{truth}$ can be estimated by Eq. (6.30g) if the worker weights $W = \widehat{W}\cup\widetilde{W}$ are fixed.

To address the high-dimensional challenge of directly solving the sub-problem (O_A^U) since the decision variable $\mathcal{F}_a = \{F_{a,n} \mid \forall o_n \in O\}$ forms a set of functions in $|O|$ dimensions, we decompose (O_A^U) into $|O|$ separate optimization problems, each focusing on an individual $F_{a,n}$. In this context, the original objective in Eq. (6.30a) can be rewritten as

$$\sum_{o_n\in O} \max_{\widetilde{x}_n^m\sim F_a^n} \left(E[x_{b,n}^{truth}] - E[x_{a,n}^{truth}]\right)^2. \tag{6.31}$$

We then allocate a specific disguise level η_n and stealth level ζ_n for each object o_n, ensuring that $\sum_{o_n\in O}\eta_n = \eta$ (with η_n representing the level of disguise introduced by DP noise for object o_n) and $\sum_{o_n\in O}\zeta_n = \zeta$ (where ζ_n denotes the level of stealth required to avoid defenders for object o_n). As a result, the constraints in Eqs. (6.30b) and (6.30c) are relaxed to

$$D_{KL}(F_{a,n} \| F_{b,n}) \leq \eta_n, \tag{6.32}$$

$$\int_{-\infty}^{+\infty} F_{a,n}(y) \cdot \log \frac{F_{a,n}^*(y)}{F_{b,n}(y)} dy \leq \zeta_n. \tag{6.33}$$

Based on these reformulations, the upper-level problem (O_A^U) is relaxed as follows,

$$(O_A^{U,o_n}): \max_{\widetilde{x}_n^m \sim F_{a,n}} \left(E[x_{b,n}^{truth}] - E[x_{a,n}^{truth}]\right)^2, \tag{6.34a}$$

$$\text{s.t. } D_{KL}(F_{a,n} \| F_{b,n}) \leq \eta_n, \tag{6.34b}$$

$$\int_{-\infty}^{+\infty} F_{a,n}(y) \cdot \log \frac{F_{a,n}^*(y)}{F_{b,n}(y)} dy \leq \zeta_n, \tag{6.34c}$$

$$\int_{-\infty}^{+\infty} F_{a,n}(y) dy = 1, \tag{6.34d}$$

$$E[x_{b,n}^{truth}] = \int_{-\infty}^{+\infty} y \cdot F_{b,n}(y) dy, \tag{6.34e}$$

$$E[x_{a,n}^{truth}] = \int_{-\infty}^{+\infty} y \cdot F_{a,n}(y) dy, \tag{6.34f}$$

$$x_{a,n}^{truth} = (\sum_{\widehat{u}_m \in \widehat{U}} \widehat{w}_m \cdot \widehat{x}_n^m + \sum_{\widetilde{u}_m \in \widetilde{U}} \widetilde{w}_m \cdot \widetilde{x}_n^m)/(\sum_{\widehat{u}_m \in \widehat{U}} \widehat{w}_m + \sum_{\widetilde{u}_m \in \widetilde{U}} \widetilde{w}_m). \tag{6.34g}$$

Since the problem (O_A^{U,o_n}) is a functional optimization problem, where $F_{a,n}$ is a density function [44], we solve it using variational methods [45]. The main result is presented in Theorem 6.3.

Theorem 6.3 *Given the current defense strategy $F_{a,n}^*$ and the normal behavior $F_{b,n}$, the optimal solution of problem (O_A^{U,o_n}) (i.e., the attack strategy for object o_n) is*

$$F_{a,n} := \operatorname*{arg\,max}_{F_{a,n} \in \{F_{a,n}^{min}, F_{a,n}^{max}\}} (E[x_{b,n}^{truth}] - E[x_{a,n}^{truth}])^2, \tag{6.35}$$

where $x_{a,n}^{truth}$ is estimated by Eq. (6.34g), $E[x_{a,n}^{truth}]$ is calculated by Eq. (6.34f), and

$$F_{a,n}^{min} = \frac{F_{b,n}(y) \cdot \left(\frac{F_{b,n}(y)}{F_{a,n}^{*}(y)}\right)^{\frac{\lambda_2}{\lambda_1}} \cdot \exp(-\frac{\theta_1}{\lambda_1} y - F_{b,n}(y))}{\int_{-\infty}^{+\infty} F_{b,n}(y) \cdot \left(\frac{F_{b,n}(y)}{F_{a,n}^{*}(y)}\right)^{\frac{\lambda_2}{\lambda_1}} \cdot \exp(-\frac{\theta_1}{\lambda_1} y - F_{b,n}(y)) dy}, \tag{6.36}$$

$$F_{a,n}^{max} = \frac{F_{b,n}(y) \cdot \left(\frac{F_{b,n}(y)}{F_{a,n}^{*}(y)}\right)^{\frac{\lambda_2}{\lambda_1}} \cdot \exp(\frac{\theta_1}{\lambda_1} y - F_{b,n}(y))}{\int_{-\infty}^{+\infty} F_{b,n}(y) \cdot \left(\frac{F_{b,n}(y)}{F_{a,n}^{*}(y)}\right)^{\frac{\lambda_2}{\lambda_1}} \cdot \exp(\frac{\theta_1}{\lambda_1} y - F_{b,n}(y)) dy}. \tag{6.37}$$

The Lagrange multipliers λ_1 *and* λ_2 *are the solutions of* $D_{KL}(F_{a,n} \| F_{b,n}) = \eta_n$ *and* $\int_{-\infty}^{+\infty} F_{a,n}(y) \cdot \log \frac{F_{a,n}^{*}(y)}{F_{b,n}(y)} dy = \zeta_n$.

Proof The object (6.34a) can be transformed into

$$(O_{A,max}^{U,o_n}) : \max_{\widetilde{x}_n^m \sim F_{a,n}} E[x_{a,n}^{truth}] \quad \text{or} \quad (O_{A,min}^{U,o_n}) : \min_{\widetilde{x}_n^m \sim F_{a,n}} E[x_{a,n}^{truth}]. \tag{6.38}$$

When $(E[x_{b,n}^{truth}] - \arg\min E[x_{a,n}^{truth}])^2 \geq (E[x_{a,n}^{truth}] - \arg\max E[x_{a,n}^{truth}])^2$, the optimal solutions of problem (O_A^{U,o_n}) and problem $(O_{A,min}^{U,o_n})$ are equivalent; otherwise the optimal solutions of problem (O_A^{U,o_n}) and problem $(O_{A,max}^{U,o_n})$ are equal.

According to variational methods, we give the small variation of the optimal attack strategy $F_{a,n}$, i.e.,

$$g(y, \alpha) = F_{a,n}(y) + \alpha \cdot l(y), \tag{6.39}$$

where α is a small variable and $l(y)$ is a unknown function. By replacing $F_{a,n}(y)$ with $g(y, \alpha)$ in the problem $(O_{A,min}^{U,o_n})$, its Lagrange function can be formulated as

$$\begin{aligned} L_{\boldsymbol{\lambda}}(y, \alpha) = E[x_{a,n}^{truth}] &+ \lambda_1 \cdot \left\{ \int_{-\infty}^{+\infty} g(y, \alpha) \cdot \log \frac{g(y, \alpha)}{F_{b,n}(y)} dy - \eta_n \right\} \\ &+ \lambda_2 \cdot \left\{ \int_{-\infty}^{+\infty} g(y, \alpha) \cdot \log \frac{F_{a,n}^{*}(y)}{F_{b,n}(y)} dy - \zeta_n \right\} \\ &+ \lambda_3 \cdot \left\{ \int_{-\infty}^{+\infty} g(y, \alpha) dy - 1 \right\}, \end{aligned} \tag{6.40}$$

where $\boldsymbol{\lambda} = [\lambda_1, \lambda_2, \lambda_3]$ is Lagrange multiplier. Let

$$\theta_0 := \frac{\sum_{\widehat{u}_m \in \widehat{U}} \widehat{w}_m \cdot \widehat{x}_n^m}{\sum_{\widehat{w}_m \in \widehat{U}} \widehat{w}_m + \sum_{\widetilde{u} \in \widetilde{U}} \widetilde{w}_m}, \tag{6.41}$$

$$\theta_1 := \frac{\sum_{\widetilde{u} \in \widetilde{U}} \widetilde{w}_m}{\sum_{\widehat{w}_m \in \widehat{U}} \widehat{w}_m + \sum_{\widetilde{u} \in \widetilde{U}} \widetilde{w}_m}. \tag{6.42}$$

Powered by the constraint (6.34f), we can estimate the expectation $E[x_{a,n}^{truth}]$ by

$$E[x_{a,n}^{truth}] = \theta_0 + \theta_1 \cdot \int_{-\infty}^{+\infty} y \cdot g(y, \alpha) dy, \ \forall o_n \in O, \tag{6.43}$$

since $E[\widetilde{x}_n^{m_1}] = E[\widetilde{x}_n^{m_2}], \ \forall \widetilde{u}_{m_1}, \widetilde{u}_{m_2} \in \widetilde{U}$.

Then, we take the derivative of Lagrange function $L_{\boldsymbol{\lambda}}(\alpha)$ with respect to variable α and set $\alpha = 0$ for all possible $l(y)$, i.e.,

$$\left.\frac{\partial L_{\boldsymbol{\lambda}}(y, \alpha)}{\partial \alpha}\right|_{\alpha=0} = 0, \quad \forall l(y). \tag{6.44}$$

Due to the inherent difficulty in directly computing the partial derivative of $L_{\boldsymbol{\lambda}}(y, \alpha)$, we interchange the order of operations between the derivative and integral. Let

$$\begin{aligned} G(y, \alpha) :=& \theta_1 \cdot y \cdot g(y, \alpha) + \lambda_1 \cdot g(y, \alpha) \cdot \log \frac{g(y, \alpha)}{F_{b,n}(y)} \\ &+ \lambda_2 \cdot g(y, \alpha) \cdot \log \frac{F_{a,n}^*(y)}{F_{b,n}(y)} + \lambda_3 \cdot g(y, \alpha), \end{aligned} \tag{6.45}$$

we have

$$\frac{\partial L_{\boldsymbol{\lambda}}(y, \alpha)}{\partial \alpha} = \int_{-\infty}^{+\infty} \frac{\partial G(y, \alpha)}{\partial \alpha} dy. \tag{6.46}$$

Since $\frac{\partial G(y,\alpha)}{\partial \alpha}$ is not an odd function, we have

$$\left.\frac{\partial L_{\boldsymbol{\lambda}}(y, \alpha)}{\partial \alpha}\right|_{\alpha=0} = 0 \Leftrightarrow \left.\frac{\partial G(y, \alpha)}{\partial \alpha}\right|_{\alpha=0} = 0, \ \forall l(y), \tag{6.47}$$

which leads to

$$\theta_1 \cdot y + \lambda_1 \cdot \log \frac{F_{a,n}(y)}{F_{b,n}(y)} + \lambda_1 \cdot F_{b,n}(y) + \lambda_2 \cdot \log \frac{F_{a,n}^*(y)}{F_{b,n}(y)} + \lambda_3 = 0. \tag{6.48}$$

Through solving Eq. (6.48), we obtain the optimal solution $F_{a,n}^{min}(y)$ of problem $(O_{A,min}^{U,o_n})$ in terms of the Lagrange multipliers $\boldsymbol{\lambda} = [\lambda_1, \lambda_2, \lambda_3]$, i.e.,

$$F_{a,n}^{min}(y) = F_{b,n}(y) \cdot \left(\frac{F_{b,n}(y)}{F_{a,n}^*(y)}\right)^{\frac{\lambda_2}{\lambda_1}} \cdot \exp(-(\frac{\theta_1}{\lambda_1} y + F_{b,n}(y) + \frac{\lambda_3}{\lambda_1})). \tag{6.49}$$

The Lagrange multiplier λ_3 can be solved by $\int_{-\infty}^{+\infty} F_{a,n}^{min}(y)dy = 1$ (constraint (6.34f), leading to

$$\exp(\frac{\lambda_3}{\lambda_1}) = \int_{-\infty}^{+\infty} F_{b,n}(y) \cdot \left(\frac{F_{b,n}(y)}{F_{a,n}^*(y)}\right)^{\frac{\lambda_2}{\lambda_1}} \cdot \exp(-\frac{\theta_1}{\lambda_1}y - F_{b,n}(y))dy. \tag{6.50}$$

As a result, we have

$$F_{a,n}^{min}(y) = \frac{F_{b,n}(y) \cdot \left(\frac{F_{b,n}(y)}{F_{a,n}^*(y)}\right)^{\frac{\lambda_2}{\lambda_1}} \cdot \exp(-\frac{\theta_1}{\lambda_1}y - F_{b,n}(y))}{\int_{-\infty}^{+\infty} F_{b,n}(y) \cdot \left(\frac{F_{b,n}(y)}{F_{a,n}^*(y)}\right)^{\frac{\lambda_2}{\lambda_1}} \cdot \exp(-\frac{\theta_1}{\lambda_1}y - F_{b,n}(y))dy}. \tag{6.51}$$

Similarly, the optimal solution of problem $(O_{A,max}^{U,o_n})$ is

$$F_{a,n}^{max}(y) = \frac{F_{b,n}(y) \cdot \left(\frac{F_{b,n}(y)}{F_{a,n}^*(y)}\right)^{\frac{\lambda_2}{\lambda_1}} \cdot \exp(\frac{\theta_1}{\lambda_1}y - F_{b,n}(y))}{\int_{-\infty}^{+\infty} F_{b,n}(y) \cdot \left(\frac{F_{b,n}(y)}{F_{a,n}^*(y)}\right)^{\frac{\lambda_2}{\lambda_1}} \cdot \exp(\frac{\theta_1}{\lambda_1}y - F_{b,n}(y))dy}. \tag{6.52}$$

The Lagrange multipliers λ_1 and λ_2 in Eq. (6.51) (or Eq. (6.52)) can be found numerically for given $F_{b,n}(y)$ and $F_{a,n}^*$ by solving the following nonlinear equations:

$$D_{KL}(F_{a,n} \| F_{b,n}) = \eta_n, \tag{6.53}$$

$$\int_{-\infty}^{+\infty} F_{a,n}(y) \cdot \log \frac{F_{a,n}^*(y)}{F_{b,n}(y)} dy = \zeta_n. \tag{6.54}$$

In summary, we have obtained the optimal solution of problem (O_A^{U,o_n}). □

Lower-Level Sub-problem With the attack strategy fixed as $\mathcal{F}_a^* = \{F_{a,n}^* \mid \forall o_n \in O\}$ (refer to Sect. 6.2.2), the sub-problem in Eq. (6.29g) is addressed by iteratively performing the following two steps:

- Keeping the weights for normal workers $\widehat{W} = \{\widehat{w}_m \mid \forall \widehat{u}_m \in \widehat{U}\}$ and for malicious workers $\widetilde{W} = \{\widetilde{w}_m \mid \forall \widetilde{u}_m \in \widetilde{U}\}$ constant, compute the estimated truth $X_a^{truth} = \{x_{a,n}^{truth} \mid \forall o_n \in O\}$ as follows

$$x_{a,n}^{truth} = \frac{\sum_{\widehat{u}_m \in \widehat{U}} \widehat{w}_m \cdot \widehat{x}_n^m + \sum_{\widetilde{u}_m \in \widetilde{U}} \widetilde{w}_m \cdot \widetilde{x}_n^m}{\sum_{\widehat{u}_m \in \widehat{U}} \widehat{w}_m + \sum_{\widetilde{u}_m \in \widetilde{U}} \widetilde{w}_m}; \tag{6.55}$$

- Once $x_{a,n}^{truth}$ is obtained, update the weights for both normal and malicious workers using

$$\widehat{w}_m = \log \frac{\sum_{o_n \in O} \left(\sum_{\widehat{u}_m \in \widehat{U}} (\widehat{x}_n^m - x_{a,n}^{truth})^2 + \sum_{\widetilde{u}_m \in \widetilde{U}} (\widetilde{x}_n^m - x_{a,n}^{truth})^2 \right)}{\sum_{o_n \in O} (\widehat{x}_n^m - x_{a,n}^{truth})^2}, \tag{6.56}$$

$$\widetilde{w}_m = \log \frac{\sum_{o_n \in O} \left(\sum_{\widehat{u}_m \in \widehat{U}} (\widehat{x}_n^m - x_{a,n}^{truth})^2 + \sum_{\widetilde{u}_m \in \widetilde{U}} (\widetilde{x}_n^m - x_{a,n}^{truth})^2 \right)}{\sum_{o_n \in O} (\widehat{x}_n^m - x_{a,n}^{truth})^2} \tag{6.57}$$

Building on this process, Algorithm 12 presents an attack strategy that fuses variational techniques with an alternating optimization approach to solve (O_A). The algorithm begins by initializing the poisoning strategy $\mathcal{F}_a$ with $\mathcal{F}_b$ (i.e., $F_{a,n} = F_{b,n}$ for every $o_n \in O$) as indicated in line 1. In each iteration, it alternates between solving the upper-level problem (O_A^U) and the lower-level sub-problem from Eq. (6.29g) until convergence to the current attack strategy $\mathcal{F}_a$. Furthermore, the computational complexity of this algorithm is $O(|O| \times |U|)$.

Algorithm 13 Local minimax point calculating algorithm

Input: ϖ: the threshold for convergence; $\vartheta > 0$: the constant defined in Lemma 6.1; $0 < \rho < 1$: the penalty constant;
Output: $(\mathcal{F}_a^{*,\star}, \mathcal{F}_a^{\star})$: the local minimax point.

```
for k = 0, 1 do
    F_a^{*,k} = {F_{a,n}^{*,k} | ∀o_n ∈ O} ← Obtain current defense strategy through Algorithm 11;
    F_a^k = {F_{a,n}^k | ∀o_n ∈ O} ← Obtain current attack strategy using F_a^{*,k} through Algorithm 12;
end for
while ||F_a^{*,k} − F_a^{*,k−1}|| + ||F_a^k − F_a^{k−1}|| > ϖ do
    while ||F_a^k − F_a^{k−1}|| > ϑ do
        F_a^{*,k+1} ← Obtain current defense strategy through Algorithm 11;
        F_a^{k+1} ← Obtain current attack strategy using F_a^{*,k+1} through Algorithm 12;
        k ← k + 1;
    end while
    ϑ ← ρ · ϑ;
    k ← k + 1;
end while
F_a^{*,★} ← F_a^{*,k};   F_a^★ ← F_a^k;
```

6.4.3 Local Minimax Point of Defenders-Attackers Interaction

Since both (O_A) and (O_D) are formulated as bi-level optimization problems [43], obtaining a global minimax solution for Eq. (6.16) is generally NP-hard [46]. Consequently, our goal is to identify a local minimax point in the iterative interaction between defenders and attackers. To achieve this, we first define local minimax points [46], a pivotal concept in minimax optimization, and then propose an efficient iterative algorithm to compute such a point.

Definition 6.10 (Local Minimax Points) A point $(\mathcal{F}_{\boldsymbol{a}}^{*,\star}, \mathcal{F}_{\boldsymbol{a}}^{\star})$ is said to be a local minimax point of the problem Eq. (6.16), if there exists $\vartheta_0 > 0$ and a function h satisfying $h(\vartheta) \to 0$ as $\vartheta \to 0$, such that for any $\vartheta \in (0, \vartheta_0]$ and any $(\mathcal{F}_a^*, \mathcal{F}_a)$ satisfying $\|\mathcal{F}_a^* - \mathcal{F}_{\boldsymbol{a}}^{*,\star}\| \leq \vartheta$ and $\|\mathcal{F}_a - \mathcal{F}_{\boldsymbol{a}}^{\star}\| \leq \vartheta$, we have

$$U(\mathcal{F}_{\boldsymbol{a}}^{*,\star}, \mathcal{F}_a) \leq U(\mathcal{F}_{\boldsymbol{a}}^{*,\star}, \mathcal{F}_{\boldsymbol{a}}^{\star}) \leq \max_{\mathcal{F}_a : \|\mathcal{F}_a - \mathcal{F}_{\boldsymbol{a}}^{\star}\| \leq h(\vartheta)} U(\mathcal{F}_a^*, \mathcal{F}_a). \tag{6.58}$$

According to Definition 6.10, Jin et al. [46] introduce a crucial lemma (i.e., Lemma 6.1) for acquiring local minimax point, and its detailed lemma is given below.

Lemma 6.1 *A point $(\mathcal{F}_{\boldsymbol{a}}^{*,\star}, \mathcal{F}_{\boldsymbol{a}}^{\star})$ is a local minimax point of the problem Eq. (6.16) if and only if $\mathcal{F}_{\boldsymbol{a}}^{\star}$ is a local maximum of function $U(\mathcal{F}_{\boldsymbol{a}}^{*,\star}, \cdot)$, and there exists $\vartheta_0 > 0$ such that $\mathcal{F}_{\boldsymbol{a}}^{*,\star}$ is a local minimum of function K_ϑ for all $\vartheta \in (0, \vartheta_0]$ where $K_\vartheta(\mathcal{F}_a^*) := \max_{\mathcal{F}_a : \|\mathcal{F}_a - \mathcal{F}_{\boldsymbol{a}}^{\star}\| \leq \vartheta} U(\mathcal{F}_a^*, \mathcal{F}_a)$.*

To further illustrate the relationship between local minimax point and Nash equilibrium, we give the definition of local Nash equilibrium and then showcase an important lemma (i.e., Lemma 6.2) proposed by Jin et al. in [46].

Definition 6.11 (Local Nash Equilibrium) A point $(\mathcal{F}_{\boldsymbol{a}}^{*,\star}, \mathcal{F}_{\boldsymbol{a}}^{\star})$ is said to be a local Nash equilibrium of the problem Eq. (6.16) if and only if there exists $\vartheta > 0$ such that for any $(\mathcal{F}_a^*, \mathcal{F}_a)$ satisfying $\|\mathcal{F}_a^* - \mathcal{F}_{\boldsymbol{a}}^{*,\star}\| \leq \vartheta$ and $\|\mathcal{F}_a - \mathcal{F}_{\boldsymbol{a}}^{\star}\| \leq \vartheta$, we have

$$U(\mathcal{F}_{\boldsymbol{a}}^{*,\star}, \mathcal{F}_a) \leq U(\mathcal{F}_{\boldsymbol{a}}^{*,\star}, \mathcal{F}_{\boldsymbol{a}}^{\star}) \leq U(\mathcal{F}_a^*, \mathcal{F}_{\boldsymbol{a}}^{\star}). \tag{6.59}$$

Lemma 6.2 *Any local Nash equilibrium is a local minimax point.*

Based on Lemma 6.1, we introduce an algorithm (Algorithm 13) to compute a local minimax point for the game described in Sect. 6.3.1. In each iteration, the algorithm alternates between solving (O_D) and (O_A); that is, it sequentially executes Algorithm 11 (line 7) and Algorithm 12 (line 8). Next, it checks whether the condition $\|\mathcal{F}_a - \mathcal{F}_{\boldsymbol{a}}^{\star}\| \leq \vartheta$ (as defined in Lemma 6.1) holds (line 6). If the condition is satisfied, the threshold ϑ is reduced (line 11) and the algorithm then evaluates whether the overall convergence criteria have been met (line 5). By iteratively performing these steps, the algorithm successfully finds the local minimax point for the defender-attacker interaction.

6.5 Computational Complexity of Attacks and Defenses

We present a comprehensive analysis of the computational complexity for both the attack and defense methods. The defense algorithm (Algorithm 11) starts with an initialization step, and in each iteration, it executes crossover, mutation, fitness evaluation, and simulated annealing. Their individual complexities are $O(|O| \times r \times SN)$ for initialization, $O(|O| \times |U|)$ for crossover, $O(SN \times |r| \times |O|)$ for mutation, $O(SN \times |O|)$ for fitness evaluation, and $O(SN \times |O| \times |U|)$ for simulated annealing. As a result, the overall complexity of Algorithm 11 is $O(G_{max} \times SN \times |O| \times |U|)$. Additionally, the attack strategy algorithm (Algorithm 12) has a complexity of $O(|O| \times |U|)$. Combining these, the total computational cost of the proposed Stackelberg-game-based defense approach (Algorithm 13) is $O(G_{max} \times SN \times |O| \times |U|)$.

6.6 Performance Evaluation

The proposed game-based defense approach allows us to develop an effective countermeasure against potent data poisoning attacks. In this section, we first assess the performance of our proposed potent attack along with several existing methods on DP-based privacy-preserving crowdsensing systems, and then evaluate the effectiveness of our defense strategy in mitigating these threats.

6.6.1 Simulation Settings

Datasets Our experiments utilize two datasets: SynData (a synthetic dataset) and Emotion (a real-world dataset). Below is an overview of each.

- *SynData Dataset.* The SynData dataset comprises 200,000 records generated by 500 workers for 400 objects. The sensory data follow a normal distribution: $\widehat{x}_n^m \sim N(\widehat{\mu}_n, \widehat{\sigma}_m^2), \forall o_n \in O, \forall \widehat{u}_m \in \widehat{U}$, where $\widehat{\mu}_n$ represents the true value of object o_n, and $\widehat{\sigma}_m^2$ indicates the reliability of worker $\widehat{u}_m$. Following [5, 15], $\widehat{\mu}_n$ is sampled from $U(20, 30)$ and $\widehat{\sigma}_m^2$ from $U(0, 30)$.
- *Emotion Dataset [47].* The Emotion dataset, derived from real-world data, contains 7000 records collected from 30 workers across 700 objects. Each entry quantifies the intensity of emotions (e.g., anger, disgust, and fear) in text, with values ranging from -100 to 100. Table 6.1 presents two randomly selected samples.

Comparisons of Data Poisoning Attacks To evaluate the proposed game-based defense approach, we conducted experiments using five distinct data poisoning

Table 6.1 Sample records of "anger" in the emotion dataset

WorkerID	ObjectID	Response	Ground-truth
A1AVJRFM6L0RN8	594	25	37
ADAGUJNWMEPT6	594	80	37

attacks. The tests were carried out on a laptop equipped with an Intel i7-8750H 2.20 GHz CPU and 16GB of memory. Below, we provide an overview of these attacks.

- *Disguise-based Data Poisoning Attack (DDPA) [5].* DDPA is a state-of-the-art data poisoning attack against DP-based privacy-preserving crowdsensing systems, leveraging DP noise to mask its malicious behavior.
- *Optimization-based Attack (OA) [15].* This attack is formulated as an optimization problem, aiming to maximize the deviation introduced by poisoning. Moreover, OA is designed for crowdsensing systems that do not implement privacy protection mechanisms.
- *Substitution-based approach for black-box data poisoning attack (SubPac) [48].* This attack is formulated as a bi-level min-max optimization problem, where the outer layer focuses on identifying optimal attack strategies, while the inner layer refines data aggregation.
- *Random Attack (RA).* In RA, for each object $o_n \in O$, a malicious worker selects a value from the uniform distribution $U(x_n^{min}, x_n^{max})$ as the manipulated sensory data. Here, x_n^{min} and x_n^{max} represent the lowest and highest sensory values reported by normal workers for o_n, respectively, where $x_n^{min} = \arg\min\{\widehat{x}_n^m \mid \forall \widehat{u}_m \in \widehat{U}\}$ and $x_n^{max} = \arg\max\{\widehat{x}_n^m \mid \forall \widehat{u}_m \in \widehat{U}\}$. Thus, the malicious data follows $\widetilde{x}_n^m \sim U(x_n^{min}, x_n^{max})$ for all $o_n \in O$ and $\widehat{u}_m \in \widehat{U}$.
- *Maximum Gain Attack (MGA).* In MGA, for each object $o_n \in O$, malicious workers generate sensory data by solving the optimization problem $\max_{\widehat{x}_n^m} (\widehat{x}_n^m - x_{b,n}^{truth})^2$ to maximize the deviation from the true value. Consequently, the malicious data $\widehat{x}_n^m$ satisfies the condition $(\widetilde{x}_n^m - x_{b,n}^{truth})^2 \geq (\widehat{x}_n^m - x_{b,n}^{truth})^2, \forall \widehat{x}_n^m \in \widehat{X}$, ensuring the greatest possible discrepancy from the actual sensory data.

Comparisons of Defenses We assess our proposed defense method by comparing it with two well-known mechanisms: MWA [15] and RobTD [22].

- *MWA [15].* This approach updates worker weights using Eq. (6.13). The truth estimation involves three steps: first, sorting all uploaded sensory data and dividing them into L groups; second, calculating the truth for each group using Eq. (6.12) to obtain L estimates; finally, selecting the median of these estimates as the final truth.
- *RobTD [22].* This method computes the error bias and variance for each worker and removes those with high values to lessen the impact of malicious inputs. After filtering, worker weights and truth estimates are updated via Eq. (6.13) and Eq. (6.12), respectively.

Evaluation Metrics Based on Definition 6.2, we introduce the attack gain metric to assess the harm that malicious workers cause in crowdsensing systems. Attackers strive to maximize this metric, which reflects the deviation from the true values identified by truth discovery methods; a higher attack gain indicates a more impactful attack. Conversely, since this metric represents the loss incurred by defenders, a lower attack gain implies a stronger defense. Thus, we employ the attack gain (AG) metric to evaluate the effectiveness of both our data poisoning attack and the corresponding defense strategy.

6.6.2 Simulation Results

Effectiveness of Our Attack Strategy

We investigate the effectiveness of both our proposed and existing data poisoning strategies on CDP-based privacy-preserving crowdsensing systems in the absence of defense mechanisms. In our analysis, we adjust the proportions of known workers (see Definition 6.3) and malicious workers (see Definition 6.4) to assess the impact of these attacks. This comparative study enables us to gauge not only the strength of our attacks but also the potential vulnerabilities that CDP introduces to crowdsensing systems.

Impacts of the Privacy Budget Figure 6.2 presents the AG values achieved by our data poisoning method and competing attacks in a CDP-based privacy-preserving crowdsensing system without any defenses, with CDP enforced using the Laplace mechanism. Initially, our results show that our method can effectively exploit the noise from the Laplace mechanism to circumvent truth discovery techniques. For example, on the SynData dataset, our attack reaches an AG of 108.52 at $\varepsilon = 0.1$,

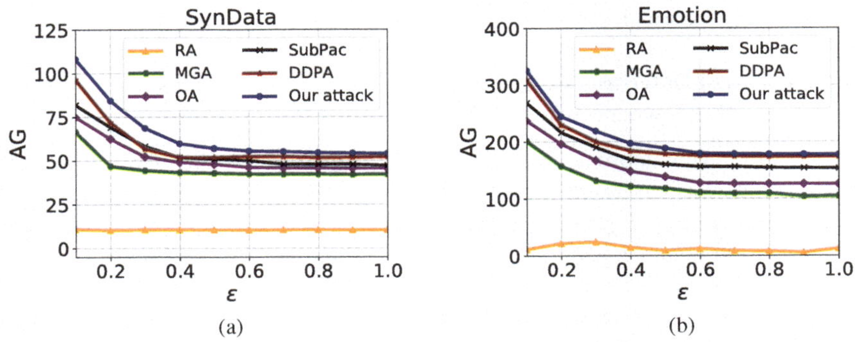

Fig. 6.2 The attack gain (AG) vs. the privacy budget ε of our data poisoning attack and competitor attacks against the ε-CDP-based privacy-preserving crowdsensing with Laplace mechanism. (**a**), (**b**) Impact of ε on attack gain

reflecting a 99.9% increase in AG as the privacy budget drops from 1.0 to 0.1. This improvement arises from our ability to use CDP noise to hide malicious actions. Additionally, our attack performs on par with DDPA, with both methods exhibiting similar AG values under the same privacy budget. This outcome is due to the lack of defenses in the CDP-based system, which allows both attacks to remain undetected.

Figure 6.3 illustrates the enhanced effectiveness of our attack on CDP-based privacy-preserving crowdsensing systems using the Gaussian mechanism, evaluated on both SynData and Emotion. With a fixed privacy budget ε and relaxation parameter δ, our attack achieves a higher attack gain (AG) compared to SubPac, OA, RA, and MGA. For example, at $\varepsilon = 1.0$ and $\delta = 0.1$, our method improves AG by 73.50% on SynData and 80.76% on Emotion. Notably, when no defense is deployed, the AG of our approach is nearly equivalent to that of DDPA. Furthermore, Fig. 6.4 demonstrates that decreasing the relaxation parameter δ further boosts our

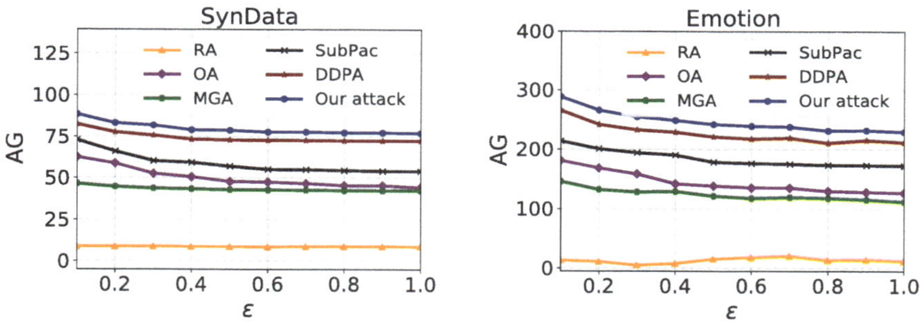

(a) (b)

Fig. 6.3 The attack gain (AG) varying with the privacy budget ε of our data poisoning attack and competitor attacks against the $(\varepsilon, 0.1)$-CDP-based privacy-preserving crowdsensing with Gaussian mechanism. (**a**), (**b**) AG vs. ε, where $\delta = 0.1$

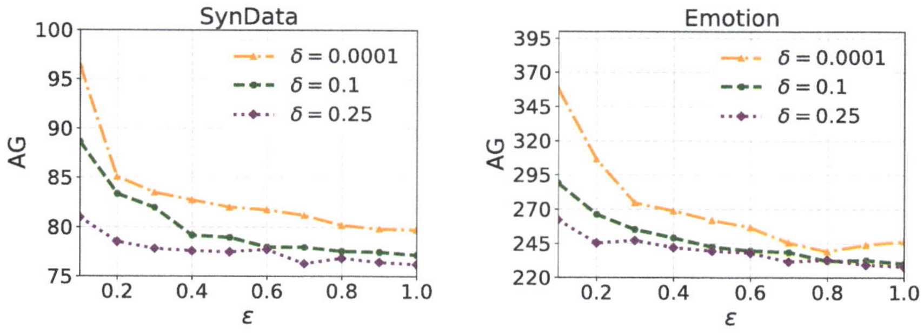

(a) (b)

Fig. 6.4 The attack gain (AG) vs. the privacy budget ε of our data poisoning attack against the (ε, δ)-CDP-based privacy-preserving crowdsensing with Gaussian mechanism, where the relaxation parameter $\delta = 0.0001, 0.1, 0.25$. (**a**), (**b**) Impact of (ε, δ) on attack gain

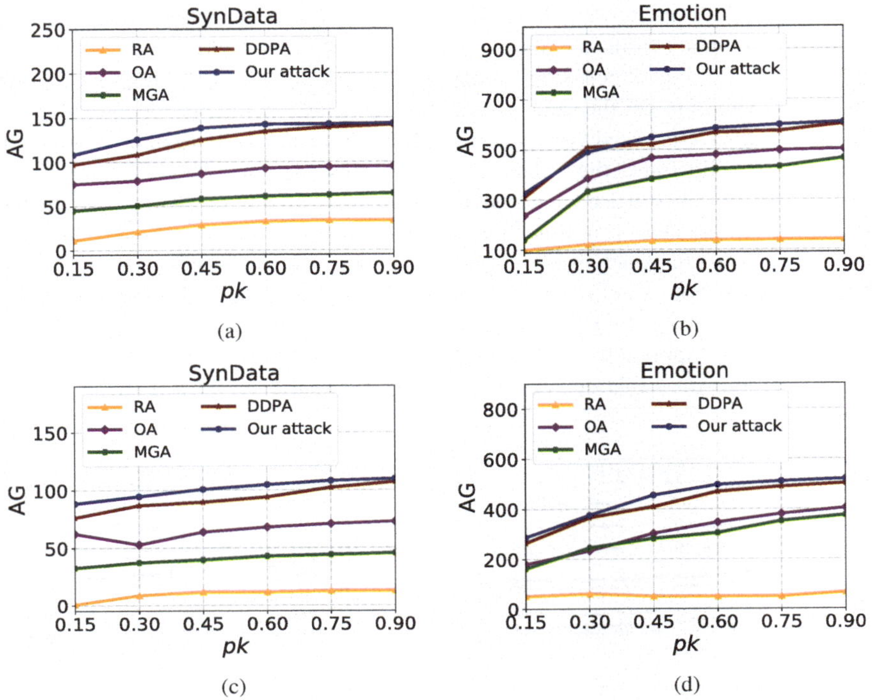

Fig. 6.5 The attack gain (AG) vs. the percentage of known workers (pk) of our data poisoning attack and competitor attacks against the CDP-based privacy-preserving crowdsensing, where CDP achieved by (**a**), (**b**) the Laplace mechanism and (**c**), (**d**) the Gaussian mechanism

attack's performance. Specifically, as δ drops from 0.25 to 0.0001 at $\varepsilon = 0.1$, the AG enhancement reaches 19.07% on SynData and 36.67% on Emotion. This improvement occurs because a lower δ enforces stronger privacy protection, thereby injecting more CDP noise into the sensory data and enabling attackers to better mask their malicious behavior.

Impacts of the Percentage of Known Workers Figure 6.5 presents the attack gain (AG) achieved by our method and competing attacks as the percentage of known workers pk varies on both SynData and Emotion datasets. The results show that, regardless of pk, our approach boosts AG by 96.81 on SynData and 226.2936 on Emotion when using the Laplace mechanism, and by 87.4503 on SynData and 234.8072 on Emotion with the Gaussian mechanism. Moreover, the AG for every attack increases as pk grows. This enhancement is due to a higher proportion of known workers enabling attackers to more accurately estimate the pre-attack truths X_b^{truth}, thus refining their attack strategies.

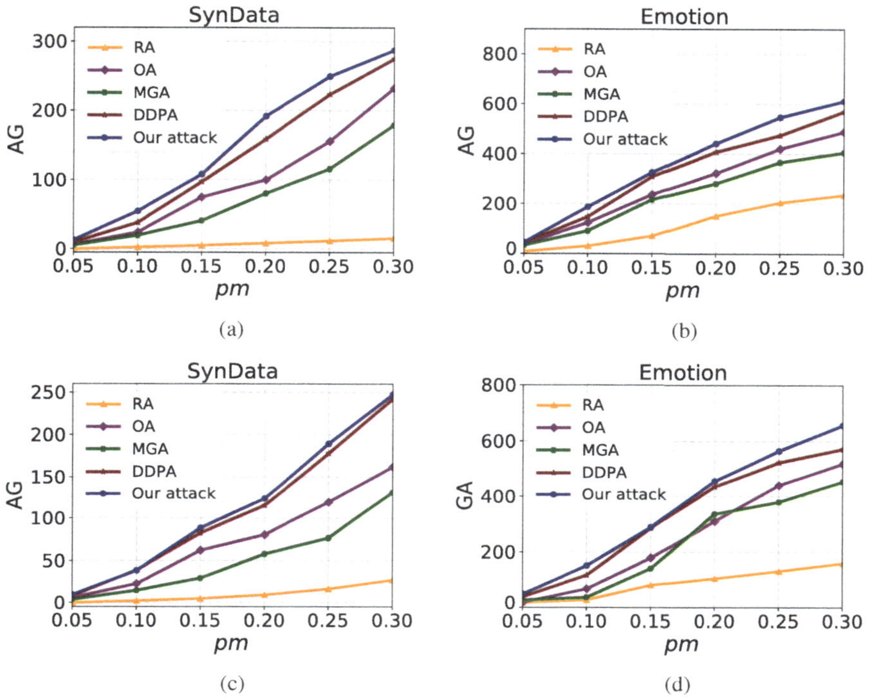

Fig. 6.6 The attack gain (AG) vs. the percentage of malicious workers (pm) of our data poisoning attack and competitor attacks against the CDP-based privacy-preserving crowdsensing, where CDP achieved by (**a**), (**b**) the Laplace mechanism and (**c**), (**d**) the Gaussian mechanism

Impacts of the Percentage of Malicious Workers Figure 6.6 illustrates how the attack gain (AG) of our approach and competing methods changes with varying percentages of malicious workers pm. As pm increases from 0.05 to 0.3, all attacks show improved performance, with AG rising by at least 274.5865 under the Laplace mechanism and 238.1736 under the Gaussian mechanism on both SynData and Emotion datasets. Notably, our attack and DDPA experience more substantial improvements as pm grows, since a higher proportion of malicious workers enhances the attackers' ability to manipulate the system successfully.

Effectiveness of Our Defense Strategy Figure 6.7 compares our defense method with competitors MWA and RobTD. Our approach consistently achieves a much larger reduction in attack gain (AG) for the same privacy budget ε. For instance, when $\varepsilon = 0.1$ under the Laplace mechanism, our defense decreases AG by 181.9455 on SynData, whereas MWA and RobTD achieve reductions of 26.3168 and 50.5390, respectively. This advantage arises because our method can detect malicious workers hidden by CDP noise, an ability that MWA and RobTD lack. Moreover, our defense's effectiveness improves as the privacy budget decreases;

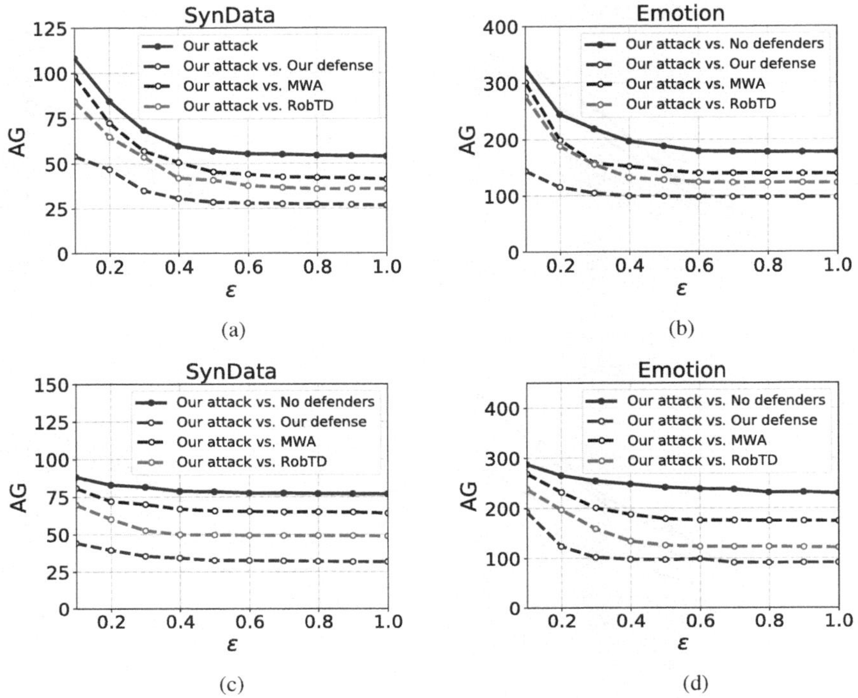

Fig. 6.7 The attack gain (AG) vs. the privacy budget ε of our defense approach and competitor defenses resist our data poisoning attack. (**a**), (**b**) the Laplace mechanism and (**c**), (**d**) the Gaussian mechanism

Table 6.2 Computational complexity of defenses

Defense strategy	Computational complexity
MWA	$O(\lvert O\rvert \times \lvert U\rvert \times \log \lvert U\rvert)$
RobTD	$O(\lvert O\rvert \times \lvert U\rvert \times \log \lvert U\rvert)$
Our	$O(G_{max} \times SN \times \lvert O\rvert \times \lvert U\rvert)$

for example, on SynData, the reduction in AG increases from 79.9844 at $\varepsilon = 1.0$ to 181.9455 at $\varepsilon = 0.1$. This is due to the lower ε introducing more CDP noise, which helps expose the attackers' activities. Additionally, Table 6.2 compares the computational complexity of our approach with that of MWA and RobTD. While our method requires slightly more computation when $\log |U| < G_{max} \times SN$, this overhead becomes negligible as the number of workers $|U|$ grows.

We further examine the robustness of our defense strategy by considering four scenarios: (DDPA vs. No defenders), (Our attack vs. No defenders), (DDPA vs. Our defense), and (Our attack vs. Our defense). In the first two cases, where either DDPA or our attack is deployed against DP-based privacy-preserving crowdsensing systems without any defenses, attackers can fully exploit their techniques, which are the most favorable conditions for adversaries. In contrast, defenders benefit from scenarios like (DDPA vs. Our defense), where attackers are unaware of the countermeasures and cannot bypass them. However, neither extreme is stable: without defenses, attackers maximize AG, while effective defense minimizes it. Ultimately, the scenario (Our attack vs. Our defense) represents a stable equilibrium, as our game-based defense approach successfully mitigates a potent data poisoning attack. Comparing these four configurations allows us to thoroughly assess the effectiveness of our defense mechanism.

Impacts of the Privacy Budget Figure 6.8 demonstrates how both attackers and defenders deploy their respective strategies under different privacy budgets. First, by comparing the scenarios (DDPA vs. No defenders) and (DDPA vs. Our defense), we observe that our proposed defense markedly diminishes the impact of DDPA, as evidenced by a notable decrease in AG for every privacy budget value. For example, at $\varepsilon = 0.1$, the AG metric falls by 80.7819 (Laplace) and 69.7213 (Gaussian) on the Syndata dataset, and by 211.6281 (Laplace) and 254.6977 (Gaussian) on the Emotion dataset. Second, an analogous comparison between (Our attack vs.

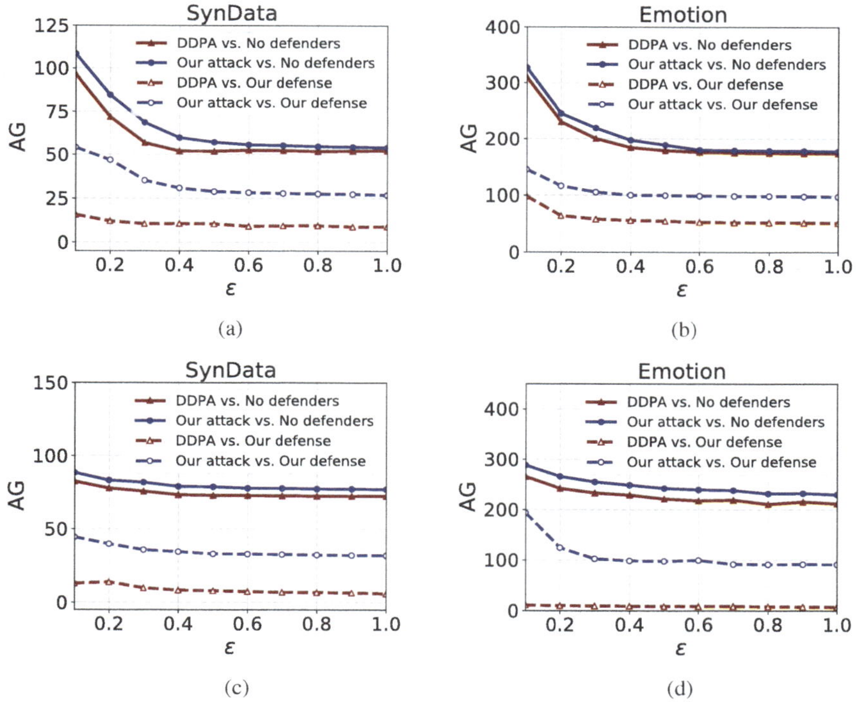

Fig. 6.8 The attack gain (AG) vs. the privacy budget ε for the attackers. The various lines depict combinations of our (○) and other existing (△) data poisoning attacks against the CDP-based privacy-preserving crowdsensing with (– –) and without (–) our defense strategy. (**a**), (**b**) the Laplace mechanism and (**c**), (**d**) the Gaussian mechanism

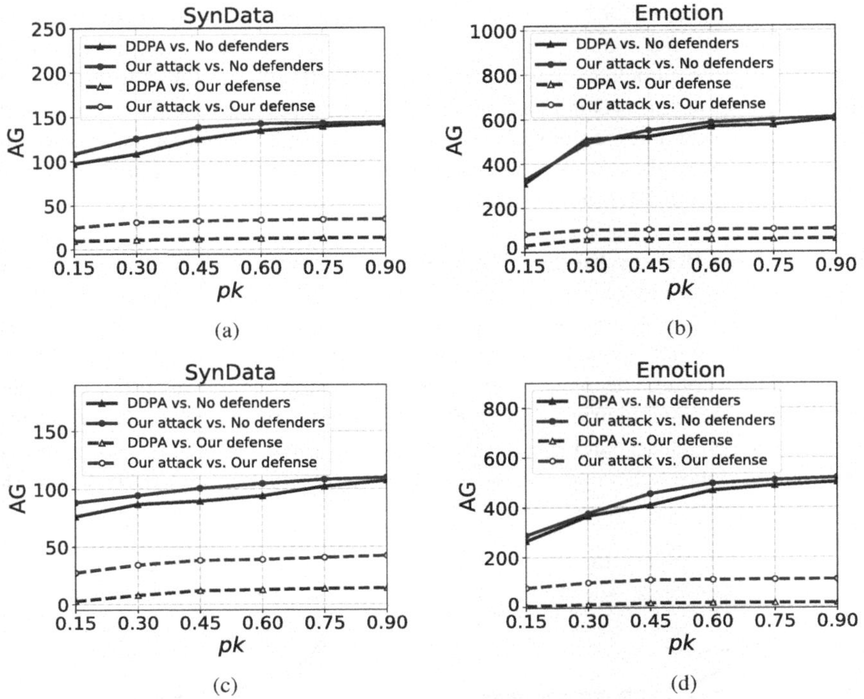

Fig. 6.9 The attack gain (AG) vs. the percentage of known workers (pk). The various lines depict combinations of our (∘) and other existing (△) data poisoning attacks against the DP-based privacy-preserving crowdsensing with (− −) and without (−) our defense strategy. (**a**), (**b**) the Laplace mechanism and (**c**), (**d**) the Gaussian mechanism

No defenders) and (Our attack vs. Our defense) shows that our defense approach effectively counters our data poisoning attack. At $\varepsilon = 0.1$, the AG reduces by 54.2027 (Laplace) and 44.0708 (Gaussian) on SynData, and by 181.9455 (Laplace) and 94.5856 (Gaussian) on Emotion. Lastly, when comparing (Our attack vs. Our defense) with (DDPA vs. Our defense), it is evident that the data poisoning attack can partly evade the defense mechanism. Specifically, our attack yields an AG increase of at least 19.9095 on Syndata and 45.8921 on Emotion relative to DDPA.

Impacts of the Percentage of Known Workers Figure 6.9 demonstrates how our defense mechanism's performance changes with varying proportions of known workers, represented by pk. Initially, by contrasting the results from (DDPA vs. No defenders) with (DDPA vs. Our defense), we find that our approach consistently counters DDPA, yielding a marked reduction in AG for all pk values. For example, when $pk = 0.15$, the Laplace mechanism on the SynData dataset shows an AG decrease of 87.5258. Similarly, by comparing (Our attack vs. No defenders) to (Our attack vs. Our defense), it becomes evident that our strategy effectively mitigates

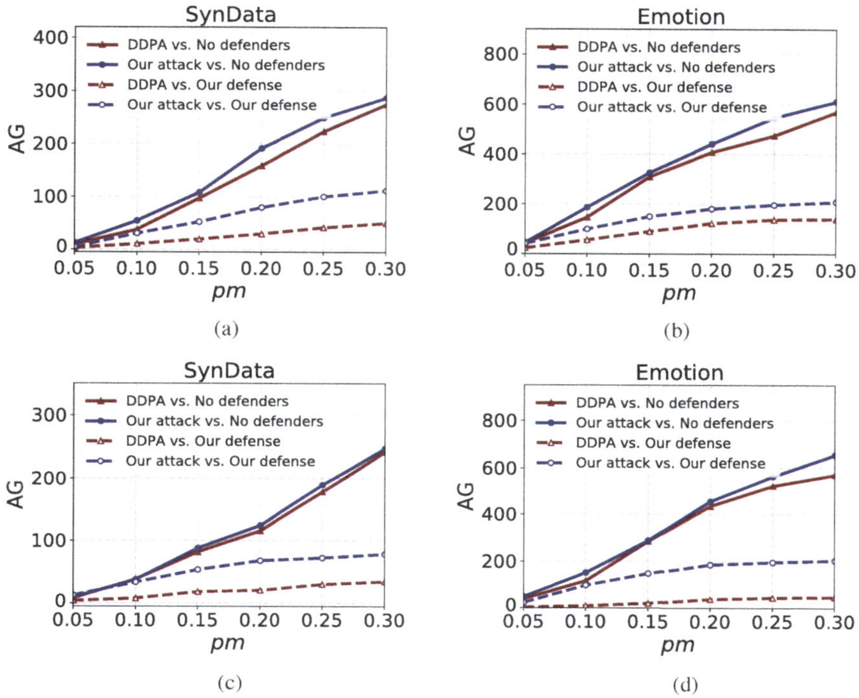

Fig. 6.10 The attack gain (AG) vs. the percentage of malicious workers (pm). The various lines depict combinations of our (∘) and other existing (△) data poisoning attacks against the CDP-based privacy-preserving crowdsensing with (– –) and without (–) our defense strategy. (**a**), (**b**) the Laplace mechanism and (**c**), (**d**) the Gaussian mechanism

poisoning attacks across any pk, with a reduction of 83.3500 in AG (Laplace) on SynData at $pk = 0.15$. Finally, examining the differences between (DDPA vs. Our defense) and (Our attack vs. Our defense) indicates that the defense's effectiveness remains robust, as AG does not fluctuate significantly with changes in pk.

Impacts of the Percentage of Malicious Workers Figure 6.10 displays how the effectiveness of our defense strategy shifts with different levels of malicious worker participation, indicated by pm. Our defense mechanism successfully counters DDPA: for any given pm, the AG value when using our defense (i.e. DDPA vs. Our defense) is substantially lower than that when no defense is applied (i.e. DDPA vs. No defenders). In the case of our poisoning attack, the pattern is similar: the AG obtained with our defense (Our attack vs. Our defense) is notably reduced compared to the scenario without any defense (Our attack vs. No defenders) for all values of pm. Additionally, although AG tends to increase gradually as pm rises, our defense maintains its robustness. Comparisons between the defense scenarios (DDPA vs.

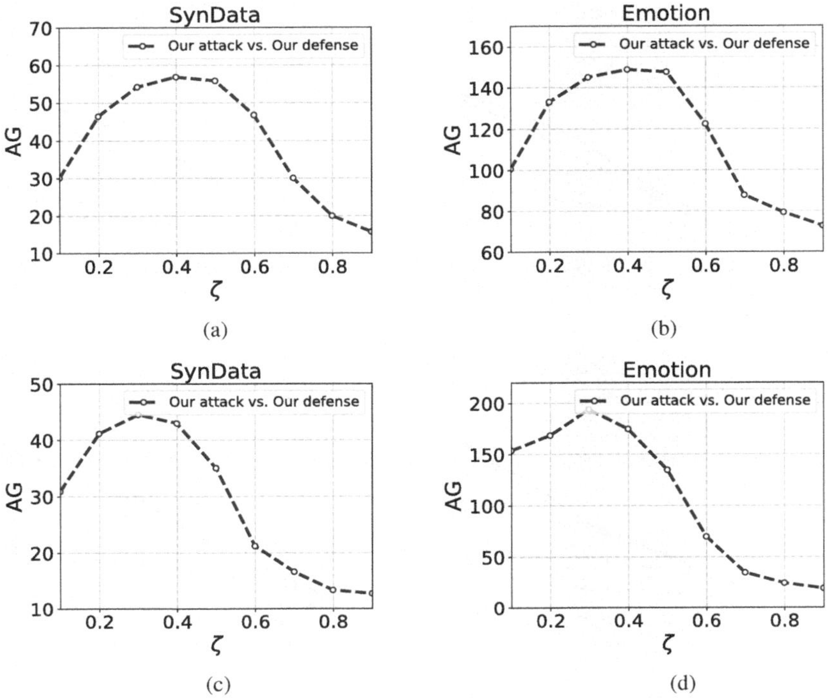

Fig. 6.11 The effectiveness of our defense strategy varies with the stealth level (ζ). The attack gain (AG) vs. the stealth level (ζ) of our data poisoning attacks against the CDP-based privacy-preserving crowdsensing with defenders, where CDP is achieved by (**a**), (**b**) the Laplace mechanism and (**c**), (**d**) Gaussian mechanism

Our defense and Our attack vs. Our defense) confirm that it effectively mitigates the bias induced by both DDPA and the poisoning attack.

Impacts of the Stealth Level Figure 6.11 illustrates how AG, resulting from our data poisoning attack on the CDP-based privacy-preserving crowdsensing system under our defense, varies with the stealth level ζ. Since ζ solely affects our attack's ability to avoid detection and does not influence other attacks, the figure only presents data for the (Our attack vs. Our defense) scenario. It can be seen that as ζ increases, AG first rises and then falls. In other words, a moderate increase in ζ initially weakens the defense, but beyond a certain point, the defense becomes more effective. This behavior is due to larger ζ values making it easier for defenders to identify malicious workers, thereby limiting the attack's impact. Conversely, a lower ζ enhances the attackers' evasion capability, lowering the expected log-likelihood ratio test, yet when ζ is too small, the overall damage inflicted by the attack is reduced.

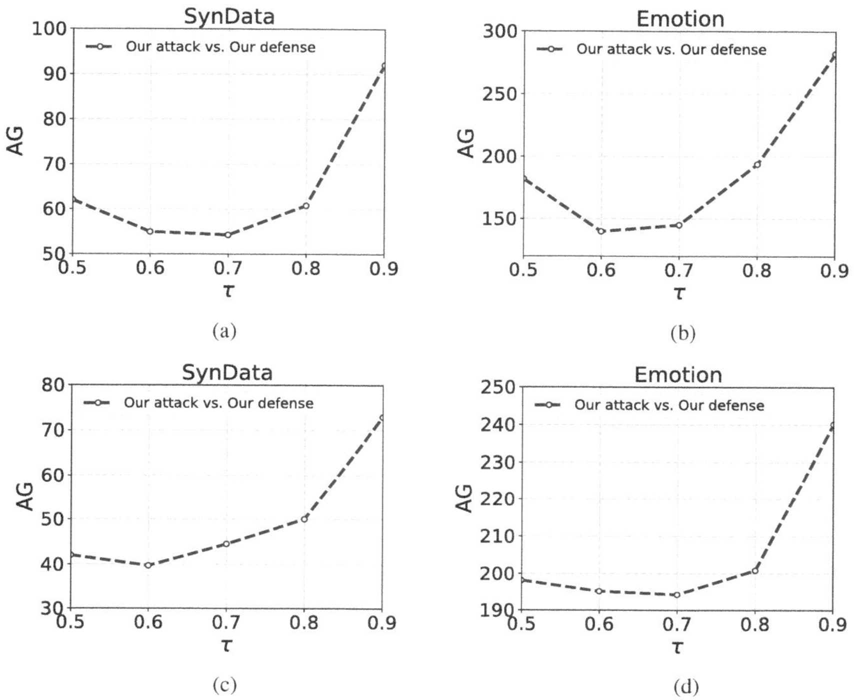

Fig. 6.12 The effectiveness of our defense strategy varies with the parameter of hypothesis testing (τ). The attack gain (AG) vs. the parameter (ζ) of our data poisoning attack against the CDP-based privacy-preserving crowdsensing with defenders, where CDP is achieved by (**a**), (**b**) the Laplace mechanism and (**c**), (**d**) Gaussian mechanism

Impacts of the Parameter of Hypothesis Testing Figure 6.12 shows how our defense strategy's performance changes with the hypothesis testing parameter, τ, which represents the likelihood that a worker is malicious. The figure focuses on the (Our attack vs. Our defense) scenario, as τ directly affects the defense's effectiveness. As τ increases, AG initially drops and then rises. This pattern suggests that our defense works best at an optimal range of τ: if τ is too high, defenders might misclassify malicious workers as benign, while if it is too low, honest workers could be mistakenly flagged as malicious.

6.7 Summary and Further Reading

We introduce a defense mechanism grounded in Stackelberg game theory to counter potent data poisoning attacks in CDP-based privacy-preserving crowdsensing systems. Our findings show that even when attackers are aware of the defense strategy, the system remains resilient against their sophisticated attacks. Moreover, this method can be readily extended to other CDP-based privacy-preserving systems such as location-based services to alleviate the detrimental effects associated with CDP. For the future work, we plan to investigate a more advanced data poisoning attack that camouflages itself within the noise generated by shuffle DP, which differs from conventional DP noise approaches since shuffle DP ensures strong privacy protection with only minimal noise addition.

References

1. Mao, X., Miao, X., He, Y., Li, X.Y., Liu, Y.: Citysee: Urban co2 monitoring with sensors. In: Proceedings of IEEE International Conference on Computer Communications, pp. 1611–1619 (2012)
2. Wang, F.Y.: Scanning the issue and beyond: Crowdsourcing for field transportation studies and services. IEEE Trans. Intell. Transp. Syst. **16**(1), 1–8 (2015)
3. Wang, Z., Li, J., Hu, J., Ren, J., Wang, Q., Li, Z., Li, Y.: Towards privacy-driven truthful incentives for mobile crowdsensing under untrusted platform. IEEE Trans. Mob. Comput. **22**(2), 1198–1212 (2023)
4. Pang, X., Wang, Z., Liu, D., Lui, J.C.S., Wang, Q., Ren, J.: Towards personalized privacy-preserving truth discovery over crowdsourced data streams. IEEE/ACM Transactions on Networking **30**(1), 327–340 (2022)
5. Li, Z., Zheng, Z., Guo, S., Guo, B., Xiao, F., Ren, K.: Disguised as privacy: Data poisoning attacks against differentially private crowdsensing systems. IEEE Trans. Mob. Comput. **22**(9), 5155–5169 (2023)
6. Zheng, Z., Li, Z., Jiang, H., Zhang, L.Y., Tu, D.: Semantic-aware privacy-preserving online location trajectory data sharing. IEEE Trans. Inf. Forensics Secur. **17**, 2256–2271 (2022)
7. Huang, C., Liu, D., Yang, A., Lu, R., Shen, X.: Multi-client secure and efficient DPF-based keyword search for cloud storage. IEEE Trans. Dependable Secur. Comput. **21**(1), 353–371 (2023)
8. Hou, J., Liu, D., Huang, C., Zhuang, W., Shen, X., Sun, R., Ying, B.: Data protection: Privacy-preserving data collection with validation. IEEE Trans. Dependable Secur. Comput. **21**(4), 3422–3438 (2023)
9. Huang, C., Wang, W., Liu, D., Lu, R., Shen, X.: Blockchain-assisted personalized car insurance with privacy preservation and fraud resistance. IEEE Trans. Veh. Technol. **72**(3), 3777–3792 (2023)
10. Dwork, C., McSherry, F., Nissim, K., Smith, A.: Calibrating noise to sensitivity in private data analysis. In: Proceedings of Theory of Cryptography, pp. 265–284 (2006)
11. Li, Y., Xiao, H., Qin, Z., Miao, C., Su, L., Gao, J., Ren, K., Ding, B.: Towards differentially private truth discovery for crowd sensing systems. In: Proceedings of IEEE International Conference on Distributed Computing Systems, pp. 1156–1166 (2020)
12. Xu, G., Li, H., Xu, S., Ren, H., Zhang, Y., Sun, J., Deng, R.H.: Catch you if you deceive me: Verifiable and privacy-aware truth discovery in crowdsensing systems. In: Proceedings of ACM Asia Conference on Computer and Communications Security, pp. 178–192 (2020)
13. Sun, P., Wang, Z., Wu, L., Feng, Y., Pang, X., Qi, H., Wang, Z.: Towards personalized privacy-preserving incentive for truth discovery in mobile crowdsensing systems. IEEE Trans. Mob. Comput. **21**(1), 352–365 (2022)
14. Li, Y., Miao, C., Su, L., Gao, J., Li, Q., Ding, B., Qin, Z., Ren, K.: An efficient two-layer mechanism for privacy-preserving truth discovery. In: Proceedings of ACM SIGKDD International Conference on Knowledge Discovery & Data Mining, pp. 1705–1714 (2018)

15. Fang, M., Sun, M., Li, Q., Gong, N.Z., Tian, J., Liu, J.: Data poisoning attacks and defenses to crowdsourcing systems. In: Proceedings of World Wide Web Conference, pp. 969–980 (2021)
16. Miao, C., Li, Q., Su, L., Huai, M., Jiang, W., Gao, J.: Attack under disguise: An intelligent data poisoning attack mechanism in crowdsourcing. In: Proceedings of World Wide Web Conference, pp. 13–22 (2018)
17. Zhang, H., Li, M.: Multi-round data poisoning attack and defense against truth discovery in crowdsensing systems. In: Proceedings of IEEE International Conference on Mobile Data Management, pp. 109–118 (2022)
18. Zhao, Y., Gong, X., Lin, F., Chen, X.: Data poisoning attacks and defenses in dynamic crowdsourcing with online data quality learning. IEEE Trans. Mob. Comput. **22**(5), 2569–2581 (2023)
19. Miao, C., Li, Q., Xiao, H., Jiang, W., Huai, M., Su, L.: Towards data poisoning attacks in crowd sensing systems. In: Proceedings of ACM International Symposium on Mobile Ad Hoc Networking and Computing, pp. 111–120 (2018)
20. Fujimoto, R., Kamiyama, N.: Poisoning attacks in crowdsensing over multiple areas. In: Proceedings of IEEE Global Communications Conference, pp. 68–73 (2022)
21. Zheng, Z., Li, Z., Huang, C., Long, S., Li, M., Shen, X.: Data poisoning attacks and defenses to ldp-based privacy-preserving crowdsensing. IEEE Trans. Dependable Secur. Comput. **21**(5), 4861–4878 (2024)
22. Huang, Z., Pan, M., Gong, Y.: Robust truth discovery against data poisoning in mobile crowdsensing. In: Proceedings of IEEE Global Communications Conference, pp. 1–6 (2019)
23. Zhang, H., Li, M., Sun, Y., Qu, G.: Robust truth discovery against multi-round data poisoning attacks. In: Proceedings of Wireless Algorithms, Systems, and Applications, pp. 258–270 (2022)
24. Giraldo, J., Cárdenas, A.A., Kantarcioglu, M., Katz, J.: Adversarial classification under differential privacy. In: Proceedings of Annual Network and Distributed System Security Symposium, pp. 1–18 (2020)
25. Cao, X., Jia, J., Gong, N.Z.: Data poisoning attacks to local differential privacy protocols. In: Proceedings of USENIX Security Symposium, pp. 947–964 (2021)
26. Cheu, A., Smith, A., Ullman, J.: Manipulation attacks in local differential privacy. In: Proceedings of IEEE Symposium on Security and Privacy, pp. 883–900 (2021)
27. Wu, Y., Cao, X., Jia, J., Gong, N.Z.: Poisoning attacks to local differential privacy protocols for Key-Value data. In: Proceedings of USENIX Security Symposium, pp. 519–536 (2022)
28. Li, X., Gong, N.Z., Li, N., Sun, W., Li, H.: Fine-grained poisoning attacks to local differential privacy protocols for mean and variance estimation. In: Proceedings of USENIX Security Symposium, pp. 1739–1756 (2023)
29. Cheu, A., Zhilyaev, M.: Differentially private histograms in the shuffle model from fake users. In: Proceedings of IEEE Symposium on Security and Privacy, pp. 440–457 (2022)
30. Imola, J., Chowdhury, A.R., Chaudhuri, K.: Robustness of locally differentially private graph analysis against poisoning. Preprint. arXiv:2210.14376 (2022)
31. Dwork, C., Roth, A.: The algorithmic foundations of differential privacy. Found. Trends Theoret. Comput. Sci. **9**(3–4), 211–407 (2014)
32. Li, Q., Li, Y., Gao, J., Zhao, B., Fan, W., Han, J.: Resolving conflicts in heterogeneous data by truth discovery and source reliability estimation. In: Proceedings of ACM SIGMOD International Conference on Management of Data, pp. 1187–1198 (2014)
33. Li, Y., Sun, H., Wang, W.H.: Towards fair truth discovery from biased crowdsourced answers. In: Proceedings of ACM SIGKDD International Conference on Knowledge Discovery & Data Mining, pp. 599–607 (2020)
34. Li, Y., Li, Q., Gao, J., Su, L., Zhao, B., Fan, W., Han, J.: Conflicts to harmony: A framework for resolving conflicts in heterogeneous data by truth discovery. IEEE Trans. Knowl. Data Eng. **28**(8), 1986–1999 (2016)
35. Giulini, S., Sanguineti, M.: Approximation schemes for functional optimization problems. J. Optim. Theory Appl. **140**, 33–54 (2009)

36. Kurkova, V., Sanguineti, M.: Comparison of worst case errors in linear and neural network approximation. IEEE Trans. Inf. Theory **48**(1), 264–275 (2002)
37. Salzer, H.E.: Hermite polynomials. J. Res. Natl. Bureau Stand. **48**(2), 111 (1952)
38. Hasan Örkcü, H.: Subset selection in multiple linear regression models: A hybrid of genetic and simulated annealing algorithms. Appl. Math. Comput. **219**(23), 11018–11028 (2013)
39. Bertsimas, D., Tsitsiklis, J.: Simulated annealing. Stat. Sci. **8**(1), 10–15 (1993)
40. Mirjalili, S., Mirjalili, S.: Genetic algorithm. In: Evolutionary Algorithms and Neural Networks: Theory and Applications, pp. 43–55 (2019)
41. Miller, B.L., Goldberg, D.E., et al.: Genetic algorithms, tournament selection, and the effects of noise. Complex Syst. **9**(3), 193–212 (1995)
42. Boche, H., Stanczak, S.: The kullback-leibler divergence and nonnegative matrices. IEEE Trans. Inf. Theory **52**(12), 5539–5545 (2006)
43. Nie, J., Wang, L., Ye, J.J.: Bilevel polynomial programs and semidefinite relaxation methods. SIAM J. Optim. **27**(3), 1728–1757 (2017)
44. Alessandri, A., Cervellera, C., Sanguineti, M.: Functional optimal estimation problems and their solution by nonlinear approximation schemes. J. Optim. Theory Appl. **134**, 445–466 (2007)
45. Struwe, M., Struwe, M.: Variational Methods, vol. 991. Springer (2000)
46. Jin, C., Netrapalli, P., Jordan, M.: What is local optimality in nonconvex-nonconcave minimax optimization? In: Proceedings of International Conference on Machine Learning, pp. 4880–4889 (2020)
47. Snow, R., O'connor, B., Jurafsky, D., Ng, A.Y.: Cheap and fast–but is it good? Evaluating non-expert annotations for natural language tasks. In: Proceedings of the Conference on Empirical Methods in Natural Language Processing, pp. 254–263 (2008)
48. Chen, P., Yang, Y., Yang, D., Sun, H., Chen, Z., Lin, P.: Black-box data poisoning attacks on crowdsourcing. In: Proceedings of International Joint Conference on Artificial Intelligence, pp. 2975–2983 (2023)

Chapter 7
Conclusions and Future Research Directions

7.1 Conclusions

This monograph primarily investigates perturbation-based privacy in crowdsensing from three dimensions: data utility, data privacy, and data poisoning. Specifically, the noise introduced by perturbation-based privacy-preserving mechanisms could degrade the quality of data-driven services. Thus, we focus on the trade-off between data privacy and data utility (a.k.a. data privacy vs. data utility) and propose location privacy preservation mechanisms, protecting users' location data privacy while providing high-quality location-based services. Additionally, since differential privacy, as a classical perturbation-based privacy notion, inadvertently facilitates the success of data poisoning attacks against crowdsensing systems, we focus on the trade-off between data privacy and data poisoning (a.k.a. data privacy vs. data poisoning) and propose optimization-based data poisoning attacks and defenses. By these defenses, we can defend against data poisoning attacks hidden behind the noise perturbation while protecting users' data privacy. The contributions of this monograph are summarized as follows:

- We have presented an overview of the evolutionary path, architecture, and applications of crowdsensing. Additionally, we have explored the motivations for achieving privacy preservation in crowdsensing, and emphasized the inherent privacy challenges from three dimensions: data utility, data privacy, and data poisoning.
- We have proposed a completeness privacy notions framework, linking three classical perturbation-based privacy notions: differential privacy, identifiability, and mutual-information privacy. This framework helps readers understand the perturbation-based privacy-preserving mechanisms and provides a fundamental theoretical analysis block for perturbation-based privacy in crowdsensing.
- We have proposed a semantic-aware trajectory privacy preservation mechanism that protects users' location data privacy and semantic privacy while preserving

Z. Zheng et al., *Perturbation Based Privacy in Crowdsensing*, Wireless Networks,
https://doi.org/10.1007/978-3-031-95052-0_7

semantic-aware data utility. To measure privacy leakage and utility loss caused by sharing perturbed location trajectories, a trajectory-level data privacy metric, a semantic privacy metric, and a data quality metric are proposed. Powered by these metrics, the privacy preservation mechanism is formulated as a multi-objective optimization problem, minimizing privacy leaks given the utility constraint. Additionally, we have theoretically proved that the mechanism can protect users' privacy and experimentally validated its effectiveness.

- We have proposed a pricing-aware location privacy preservation mechanism that provides users with high-quality ride-matching and dynamic pricing services while protecting their location data privacy. To overcome the abrupt changes of price multipliers near hotspots, an adaptive supply and demand aware grid is proposed. Based on this grid, we have put forward a dynamic pricing utility metric and a ride-matching utility metric to measure service quality losses arising from the noise perturbation. Then, the privacy preservation mechanism is formulated a multi-objective optimization problem, minimizing service quality losses given a privacy constraint. As a result, theoretical proofs and extensive experiments have shown that the mechanism is efficient in terms of data utility and data privacy.
- We have proposed data poisoning attacks and defenses based on bi-level optimization, mitigating poisoning threats caused by local differential privacy (LDP). The attacks are formulated as a bi-level optimization problem that conceals their attack behavior behind the noise perturbation introduced by LDP. By the proposed attacks, we have demonstrated that LDP, as a classical perturbation-based privacy notion, can facilitate the success of data poisoning attacks against crowdsensing. Furthermore, we formulate the defenses as a bi-level optimization problem, minimizing poisoning disruptions by identifying and then dropping workers manipulated by attackers. As a result, we can defend against poisoning attacks hidden behind LDP noise while protecting users' data privacy. In addition, extensive experiments on real-world datasets have shown the effectiveness of the proposed attacks and defenses.
- We have proposed game-theoretic-based data poisoning attacks and defenses that defend against poisoning threats arising from central differential privacy (CDP). The attacks-defends interaction is formulated as a zero-sum Stackelberg game, where the attacks aim to remain stealthy by weaponizing CDP and spying on the defenses. To find the equilibrium solution of this game, we have proposed a local minimax point calculating algorithm. The proposed attacks hidden behind the CDP noise have demonstrated that CDP poses more a serious poisoning threat to crowdsensing systems. By the defenses, we can mitigate poisoning threats caused by CDP while protecting users' data privacy. As a result, extensive experiments have shown that the defenses can accurately distinguish malicious behaviors disguised.

7.2 Future Research Directions

We have focused on perturbation-based privacy in crowdsensing and proposed a series of works across three dimensions: data utility, data privacy, and data poisoning. However, there are still several research issues that need further exploration.

Providing Stricter Location Privacy Protection for Users While Ensuring Service Quality We have focused on the balance between data utility and data privacy and proposed location privacy preservation mechanisms for local, online, and non-sporadic location sharing scenarios. However, these mechanisms do not overcome the issue of social relationship privacy leakage (e.g., friends, colleagues, and classmates) caused by correlations in different users' trajectories. The social relationship plays a crucial role in privacy protection, as attackers can infer victims' sensitive information from their friends' publicly available data. Besides, privacy preservation mechanisms in local scenarios can only observe an individual's data instead of the entire dataset. Therefore, protecting users' social relationship privacy in local, online, and non-sporadic scenarios remains a significant challenge.

Developing More Robust Data Poisoning Attacks and Defenses We have focused on the balance between data privacy and data poisoning, proposing data poisoning attacks and defenses to mitigate poisoning threats arising from the noise introduced by DP-based privacy preservation mechanisms. However, these attacks and defenses overlook the following two privacy preservation cases. First, personalized privacy preservation mechanisms have already been widely applied in crowdsensing systems to improve service quality by adding personalized noise perturbation. Second, mutual-information privacy better protects users' privacy in relational datasets (e.g., trajectory data, and grid data), overcoming privacy leaks arising from differences in correlation between the noise perturbation and the real data. Consequently, it is crucial to explore whether the noise perturbation introduced by personalized and mutual-information-based privacy preservation mechanisms facilitates the success of data poisoning attacks against crowdsensing. This would provide new insights for further mitigating poisoning threats arising from this noise.

The manufacturer's authorised representative in the EU is Springer Nature Customer Service Centre GmbH, Europaplatz 3, 69115 Heidelberg, Germany. If you have any concerns regarding our products, please contact ProductSafety@springernature.com

Printed and bound by CPI Group (UK) Ltd, Croydon, CR0 4YY
30/09/2026
02225572-0001